Mathematical Engineering

Series editors

Today, the development of high-tech systems is unthinkable without mathematical modeling and analysis of system behavior. As such, many fields in the modern engineering sciences (e.g. control engineering, communications engineering, mechanical engineering, and robotics) call for sophisticated mathematical methods in order to solve the tasks at hand.

The series Mathematical Engineering presents new or heretofore little-known methods to support engineers in finding suitable answers to their questions, presenting those methods in such manner as to make them ideally comprehensible and applicable in practice.

Therefore, the primary focus is—without neglecting mathematical accuracy—on comprehensibility and real-world applicability.

To submit a proposal or request further information, please use the PDF Proposal Form or contact directly: *Dr. Jan-Philip Schmidt, Publishing Editor (jan-philip.schmidt@springer.com).*

More information about this series at http://www.springer.com/series/8445

Marcelo Epstein

Partial Differential Equations

Mathematical Techniques for Engineers

Springer

Marcelo Epstein
Department of Mechanical
and Manufacturing Engineering
University of Calgary
Calgary, AB
Canada

ISSN 2192-4732 ISSN 2192-4740 (electronic)
Mathematical Engineering
ISBN 978-3-319-55211-8 ISBN 978-3-319-55212-5 (eBook)
DOI 10.1007/978-3-319-55212-5

Library of Congress Control Number: 2017934212

Printed on acid-free paper

This Springer imprint is published by Springer Nature
The registered company is Springer International Publishing AG
The registered company address is: Gewerbestrasse 11, 6330 Cham, Switzerland

To Ruth, Tamar and David,
my children, my builders.

Preface

A course on Mathematical Techniques for Engineers does not have a well-defined scope. What the so-called mathematical techniques needed by a well-trained engineer might be is clearly a matter of controversy. There is a direct relation between the mathematics one knows and the mathematics one is likely to use. In other words, it is often the case that the knowledge leads to the usage, rather than the other way around, as many believe. Why is it so? Because if you do not know a mathematical concept (the notion of characteristic lines, for example) you are unlikely to realize that you may need it (to describe shock waves or traffic flow, say), no matter how long you witness the phenomena (sonic booms, traffic jams) or how smart you are.

The question, therefore, is not so much what to include in a course of this nature, but why should one leave out entire mathematical sub-disciplines (graph theory, topology, functional analysis, and so on). It has become a tradition, however, in most engineering schools to expect that engineering students be exposed to at least one course on partial differential equations (PDEs), these being the backbone of various fundamental disciplines (solid mechanics, fluid mechanics, thermodynamics, electromagnetism, control of systems with distributed parameters, gravitation, etc.)

There are many excellent, even outstanding, texts and treatises on PDEs at a variety of levels. On the other hand, when a few years ago I was given the task of lecturing a graduate course on Mathematical Techniques for Engineers, a course that I am still in charge of, I found it both convenient and necessary to develop a set of class notes that would serve as a common foundation while letting each student find the book or books best suited to his or her style of learning and depth of interest. This policy has been amply rewarded by comments from the students themselves over the years. In publishing these notes, barely edited so as to preserve some of the freshness of a class environment, I hope that engineering students in other institutions may find in them some intellectual stimulus and enjoyment.

Calgary, AB, Canada
2017

Marcelo Epstein

Contents

Part I
Background

Chapter 1
Vector Fields and Ordinary Differential Equations

Although the theory of partial differential equations (PDEs) is not a mere generalization of the theory of ordinary differential equations (ODEs), there are many points of contact between both theories. An important example of this connection is provided by the theory of the single first-order PDE, to be discussed in further chapters. For this reason, the present chapter offers a brief review of some basic facts about systems of ODEs, emphasizing the geometrical interpretation of solutions as integral curves of a vector field.

1.1 Introduction

It is not an accident that one of the inventors of Calculus, Sir Isaac Newton (1642–1727), was also the creator of modern science and, in particular, of Mechanics. When we compare Kepler's (1571–1630) laws of planetary motion with Newton's $\mathbf{f} = m\mathbf{a}$, we observe a clear transition from merely descriptive laws, that apply to a small number of phenomena, to structural and explanatory laws encompassing almost universal situations, as suggested in Fig. 1.1. This feat was achieved by Newton, and later perfected by others, in formulating general physical laws in the small (differentials) and obtaining the description of any particular global phenomenon by means of a process of integration (quadrature).

In other words, Newton was the first to propose that a physical law could be formulated in terms of a system of *ordinary differential equations*. Knowledge of the *initial conditions* (position and velocity of each particle at a given time) is necessary and sufficient to predict the behaviour of the system for at least some interval of time. From this primordial example, scientists went on to look for differential equations that unlock, as it were, the secrets of Nature. When the phenomena under study involve a continuous extension in space and time one is in the presence of a *field theory*, such as is the case of Solid and Fluid Mechanics, Heat Transfer and Electromagnetism.

M. Epstein, *Partial Differential Equations*, Mathematical Engineering,
DOI 10.1007/978-3-319-55212-5_1

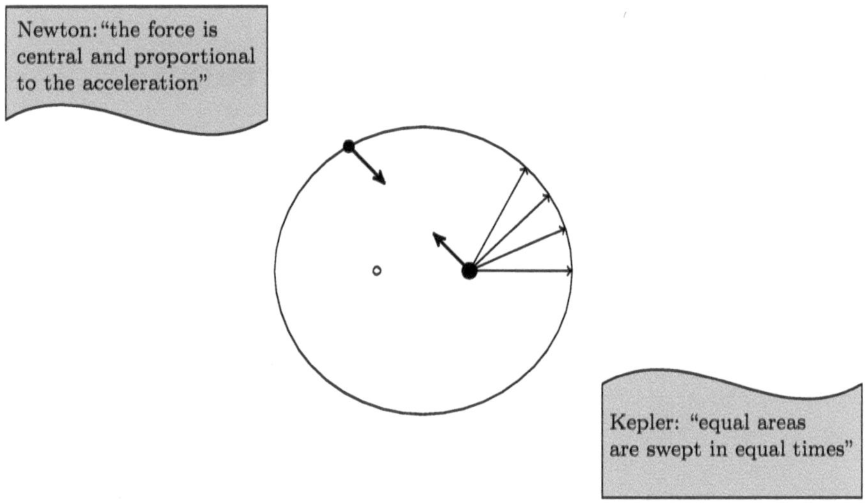

Fig. 1.1 New science from old

These phenomena can be described in terms of equations involving the fields and their partial derivatives with respect to the space and time variables, thus leading to the formulation of systems of *partial differential equations*. As we shall see in this course, and as you may know from having encountered them in applications, the analysis of these systems is not a mere generalization of the analysis of their ordinary counterparts. The theory of PDEs is a vast field of mathematics that uses the tools of various mathematical disciplines. Some of the specialized treatises are beyond the comprehension of non-specialists. Nevertheless, as with so many other mathematical areas, it is possible for engineers like us to understand the fundamental ideas at a reasonable level and to apply the results to practical situations. In fact, most of the typical differential equations themselves have their origin in engineering problems.

1.2 Curves and Surfaces in $\mathbb{R}^n$

1.2.1 Cartesian Products, Affine Spaces

We denote by $\mathbb{R}$ the set of real numbers. Recall the notion of *Cartesian product* of two sets, A and B, namely, the set $A \times B$ consisting of all ordered pairs of the form (a, b), where a belongs to A and b belongs to B. More formally,

$$A \times B = \{(a, b) \mid a \in A, b \in B\}. \tag{1.1}$$

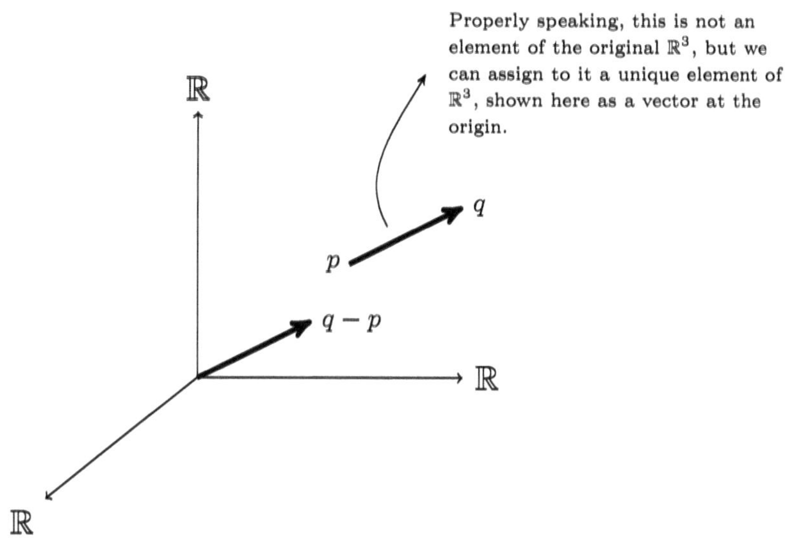

Fig. 1.2 The affine nature of $\mathbb{R}^3$

Note that the Cartesian product is not commutative. Clearly, we can consider the Cartesian product of more than two sets (assuming associativity). In this spirit we can define

$$\mathbb{R}^n = \underbrace{\mathbb{R} \times \mathbb{R} \times \cdots \times \mathbb{R}}_{n \text{ times}}. \tag{1.2}$$

Thus, $\mathbb{R}^n$ can be viewed as the set of all ordered n-tuples of real numbers. It has a natural structure of an n-dimensional vector space (by defining the vector sum and the multiplication by a scalar in the natural way).[1] The space $\mathbb{R}^n$ (or, for that matter, any vector space) can also be seen as an *affine space*. In an affine space, the elements are not vectors but *points*. To every ordered pair of points, p and q, a unique vector can be assigned in some predefined *supporting vector space*. This vector is denoted as $\overline{pq}$ or, equivalently, as the "difference" $q - p$. If the space of departure was already a vector space, we can identify this operation with the vector difference and the supporting space with the vector space itself, which is what we are going to do in the case of $\mathbb{R}^n$ (see Fig. 1.2). In this sense, we can talk about a vector *at the point* p. More precisely, however, each point of $\mathbb{R}^n$ has to be seen as carrying its own "copy" of $\mathbb{R}^n$, containing all the vectors issuing from that point. This is an important detail. For example, consider the surface of a sphere. This is clearly a 2-dimensional entity. By means of lines of latitude and longitude, we can identify a portion of this entity with $\mathbb{R}^2$, as we do in geography when drawing a map (or, more technically,

[1] The dot (or inner) product is not needed at this stage, although it is naturally available. Notice, incidentally, that the dot product is not always physically meaningful. For example, the 4-dimensional classical space-time has no natural inner product.

a *chart*) of a country or a continent. But the vectors tangent to the sphere at a point p, do not really belong to the sphere. They belong, however, to a copy of the entire $\mathbb{R}^2$ (the tangent plane to the sphere at that point). In the case in which the sphere is replaced by a plane, matters get simplified (and, at the same time, confused).

1.2.2 Curves in $\mathbb{R}^n$

Consider now a continuous *map* (that is, a continuous function)

$$\gamma : J \to \mathbb{R}^n, \tag{1.3}$$

where $J = [t_0, t_1]$ (with $t_1 > t_0$) is a (closed) interval of $\mathbb{R}$, as shown in Fig. 1.3. If we denote by t the running variable in $\mathbb{R}$, this map can be represented by n continuous functions

$$x_i = x_i(t) \qquad i = 1, \ldots, n, \tag{1.4}$$

where x_i is the running variable in the i-th copy of $\mathbb{R}$. The map γ is called a *parametrized curve* in $\mathbb{R}^n$. Since to each point $t \in J$ we assign a particular point in $\mathbb{R}^n$, we can appreciate that the above definition corresponds to the intuitive idea of a one-dimensional continuous entity in space, namely something with just one "degree of freedom". The graph of a parametrized curve (i.e., the collection of all the image points) is a *curve*. Notice that the same curve corresponds to an infinite number of

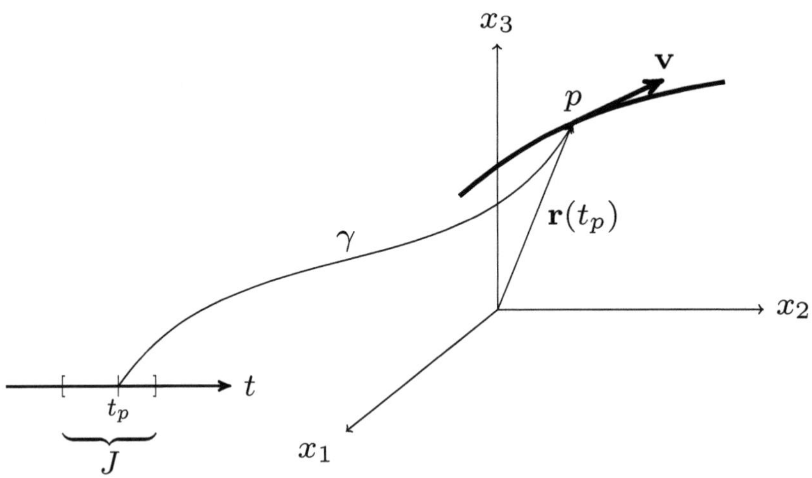

Fig. 1.3 A parametrized curve

parametrized curves. By abuse of terminology, we will usually say "curve" when actually referring to a parametrized curve.

Remark 1.1 Given a parametrized curve $\gamma : J \to \mathbb{R}^n$, a *change of parameter* is obtained by specifying any continuous strictly monotonic function $\sigma : J \to \mathbb{R}$. The composition $\gamma \circ \sigma^{-1}$ is a different parametrized curve with the same graph as the original curve γ. The simplest change of parameter is a *shift* or *translation*, given by the the function $t \mapsto r = \sigma(t) = t - a$, where a is a constant. Similarly, a *change of scale* is given by the re-parametrization $t \mapsto r = \sigma(t) = At$, where A is a positive constant. In many applications a particular choice of parameter has a clear physical meaning. In mechanics, for example, the natural parameter is time. In strength of materials, the natural parameter of (2D) Mohr's circle is the angle between normals to planes.

Remark 1.2 A continuous real function $y = f(x)$ of a single variable x can be regarded as a particular case of a parametrized curve by identifying x with the parameter. The graph of such a function is, clearly, an ordinary curve. But, contrary to the general case, this curve cannot have self-intersections or be cut by a "vertical" line in more than one point. Thus, the usual identification of plane curves with functions of one variable is incorrect.

If each of the functions $x_i(t)$ is not just continuous but also differentiable (to some order), we say that the curve is differentiable (of the same order). We say that a function is *of class* C^k if it has continuous derivatives up to and including the order k. If the curve is of class C^∞, we say that the curve is *smooth*.

It is often convenient to use a more compact vector notation[2] by introducing the so-called *position vector* $\mathbf{r}$ in $\mathbb{R}^n$, namely, the vector with components $x_1, x_2, \dots x_n$. A curve is then given by the equation

$$\mathbf{r} = \mathbf{r}(t). \tag{1.5}$$

The *tangent vector* to a differentiable curve γ at a point $p = \gamma(t_p)$ is, by definition, the vector

$$\mathbf{v} = \left. \frac{d\mathbf{r}(t)}{dt} \right|_{t=t_p}. \tag{1.6}$$

The components of the tangent vector are, accordingly, given by

$$v_i = \left. \frac{dx_i(t)}{dt} \right|_{t=t_p}. \tag{1.7}$$

[2]A luxury that we cannot afford on something like the surface of a sphere, for obvious reasons.

1.2.3 Surfaces in $\mathbb{R}^3$

A *parametrized surface* in $\mathbb{R}^3$ is the two-dimensional analog of a parametrized curve, namely a continuous map

$$\Sigma : J_1 \times J_2 \to \mathbb{R}^3, \tag{1.8}$$

where J_1 and J_2 are (closed) intervals of $\mathbb{R}$. If we denote by ξ_1 and ξ_2 the running variables in J_1 and J_2, respectively, the surface can be represented by the equation

$$\mathbf{r} = \mathbf{r}(\xi_1, \xi_2). \tag{1.9}$$

The domain of definition of the parameters ξ_1 and ξ_2 need not be limited to a rectangle, but can be any (closed) connected region in $\mathbb{R}^2$. Higher-order surfaces (or *hypersurfaces*) can be defined analogously in $\mathbb{R}^n$ by considering continuous functions $x_i = x_i(\xi_1, \ldots \xi_{n-1})$. More generally, the main object of Differential Geometry is a *differentiable manifold* of an arbitrary number of dimensions. An n-dimensional manifold can be covered with coordinate patches, each of which looks like an open set in $\mathbb{R}^n$.

Remark 1.3 A continuous real function $x_3 = f(x_1, x_2)$ of two real variables is a particular case of a parametrized surface, namely, $x_1 = \xi_1, x_2 = \xi_2, x_3 = f(\xi_1, \xi_2)$. This surface necessarily has the property that every line parallel to the x_3 axis cuts the graph of the surface at most once.

Keeping one of the coordinates (ξ_1, say) fixed and letting the other coordinate vary, we obtain a *coordinate curve* (of the ξ_2 kind, say) on the given surface. The surface can, therefore, be viewed as a one-parameter family of coordinate curves of one kind or the other. More graphically, the surface can be combed in two ways with coordinate curves, as illustrated in Fig. 1.4. In the differentiable case, a tangent vector

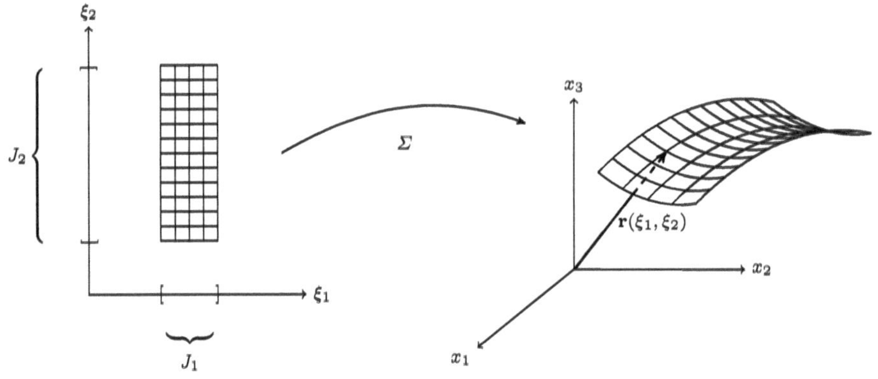

Fig. 1.4 A parametrized surface

to a coordinate curve is automatically tangent to the surface, namely, it belongs to the local tangent plane to the surface. These two tangent vectors, therefore, are given at each point of the surface by

$$\mathbf{e}_1 = \frac{\partial \mathbf{r}}{\partial \xi_1} \qquad \mathbf{e}_2 = \frac{\partial \mathbf{r}}{\partial \xi_2}. \tag{1.10}$$

They constitute a basis for the tangent plane to the surface. They are known as the *natural base vectors* associated with the given parameters ξ_1, ξ_2.

The cross product of the natural base vectors provides us, at each point, with a vector $\mathbf{m} = \mathbf{e}_1 \times \mathbf{e}_2$ perpendicular to the surface. The equation of the tangent plane at a point x_1^0, x_2^0, x_3^0 of the surface is, therefore,

$$m_1(x_1 - x_1^0) + m_2(x_2 - x_2^0) + m_3(x_3 - x_3^0) = 0, \tag{1.11}$$

where m_1, m_2, m_3 are the Cartesian components of $\mathbf{m}$.

Remark 1.4 For the particular case of a surface expressed as $x_3 = f(x_1, x_2)$, the natural base vectors (adopting x_1, x_2 as parameters) have the Cartesian components

$$\mathbf{e}_1 = \left(1, 0, \frac{\partial f}{\partial x_1}\right) \qquad \mathbf{e}_2 = \left(0, 1, \frac{\partial f}{\partial x_2}\right). \tag{1.12}$$

A vector $\mathbf{m}$ normal to the surface is given by

$$\mathbf{m} = \mathbf{e}_1 \times \mathbf{e}_2 = \left(-\frac{\partial f}{\partial x_1}, -\frac{\partial f}{\partial x_2}, 1\right), \tag{1.13}$$

and the equation of the tangent plane at (x_1^0, x_2^0, x_3^0) can be written as

$$x_3 - x_3^0 = \frac{\partial f}{\partial x_1}(x_1 - x_1^0) + \frac{\partial f}{\partial x_2}(x_2 - x_2^0). \tag{1.14}$$

1.3 The Divergence Theorem

1.3.1 The Divergence of a Vector Field

A *vector field* over a region $\mathcal{D}$ in $\mathbb{R}^3$ is an assignation of a vector $\mathbf{v}$ to each point of the region. If x_1, x_2, x_3 is a Cartesian coordinate system in $\mathcal{D}$, the vector field $\mathbf{v}$ can be given in terms of its components v_1, v_2, v_3 by means of three functions

$$v_i = v_i(x_1, x_2, x_3) \qquad i = 1, 2, 3. \tag{1.15}$$

The vector field is said to be continuous (differentiable) if each of the functions v_i is continuous (differentiable).

Given a differentiable vector field $\mathbf{v}$ we can define its *divergence* (denoted as $\nabla \cdot \mathbf{v}$ or div $\mathbf{v}$) as the scalar field given by

$$\text{div } \mathbf{v} = \frac{\partial v_1}{\partial x_1} + \frac{\partial v_2}{\partial x_2} + \frac{\partial v_3}{\partial x_3}. \tag{1.16}$$

1.3.2 The Flux of a Vector Field over an Orientable Surface

Consider an infinitesimal element of area dA at a point P in $\mathbb{R}^3$ as shown in Fig. 1.5. On the unique line orthogonal to this element we may choose either one of two unit normal vectors, one for each of the two sides of dA. Each of the two possible choices determines an *orientation* of dA. Let $\mathbf{n}$ denote one of these choices. If $\mathbf{v}$ is a vector at P, we call *the flux of v through dA with the chosen orientation* the scalar quantity $\mathbf{v} \cdot \mathbf{n}\, dA$.

A smooth surface A in $\mathbb{R}^3$ is *orientable* if an orientation $\mathbf{n}$ can be chosen smoothly on A. By this we mean that if we partition A into differential elements dA, we should be able to choose one of the two possible orientations for each element in such a way that, in the limit as the partition becomes infinitely fine, the resulting vector field $\mathbf{n}$ defined on A is smooth. Not every surface in $\mathbb{R}^3$ is orientable, as shown by the familiar example of the *Moebius band*. An orientable surface A is said to be *oriented* if one of the two possible orientations of each dA has been chosen smoothly. Given an oriented surface A and a vector field $\mathbf{v}$ defined on this surface, the *flux of* $\mathbf{v}$ *over* A is given by

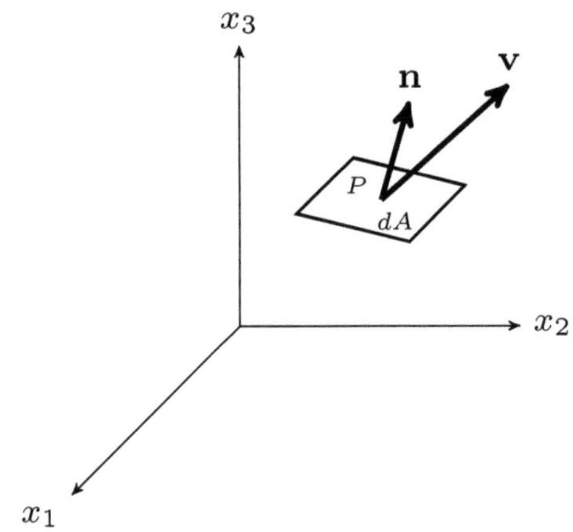

Fig. 1.5 Elementary flux $\mathbf{v} \cdot \mathbf{n}\, dA$ of a vector $\mathbf{v}$ through the oriented surface element dA

$$\text{flux}_{\mathbf{v},A} = \int_A \mathbf{v} \cdot \mathbf{n} \, dA. \tag{1.17}$$

The (Riemann) integral on the right-hand side can be regarded as the limit of the sum of the elementary fluxes as the partition is increasingly refined.

1.3.3 Statement of the Theorem

If $\mathcal{D}$ is a bounded domain in $\mathbb{R}^3$ we denote by $\partial\mathcal{D}$ its boundary.[3] The boundary of a domain is always orientable. We will systematically choose the orientation of $\partial\mathcal{D}$ as the one determined by the *exterior unit normals* $\mathbf{n}$.

Theorem 1.1 (Divergence theorem) *The integral of the divergence of a vector field on a bounded domain $\mathcal{D}$ is equal to the flux of this vector field over its boundary $\partial\mathcal{D}$, namely,*

$$\int_{\mathcal{D}} \text{div}\,\mathbf{v} \, dV = \int_{\partial\mathcal{D}} \mathbf{v} \cdot \mathbf{n} \, dA. \tag{1.18}$$

A proof of this theorem, known also as the theorem of Gauss, can be found in classical calculus books, such as [4] or [5]. A more modern and more general, yet quite accessible, formulation is presented in [6].

This fundamental result of vector calculus can be regarded as a generalization of the fundamental theorem of calculus in one independent variable. Indeed, in the one-dimensional case, if we identify the domain $\mathcal{D}$ with a segment $[a, b]$ in $\mathbb{R}$, a vector field $\mathbf{v}$ has a single component v. Moreover, the boundary $\partial\mathcal{D}$ consists of the two-element set $\{a, b\}$. The exterior unit normals at the points a and b are, respectively, the vectors with components -1 and $+1$. Thus, the divergence theorem reduces to

$$\int_a^b \frac{dv}{dx} \, dx = v(b) - v(a), \tag{1.19}$$

which reproduces the familiar result.

1.3.4 A Particular Case

Let $\phi = \phi(x_1, x_2, x_3)$ be a *scalar field*. Its *gradient*, denoted by $\textbf{grad}\phi$ or $\nabla\phi$ is the vector field with Cartesian components

[3]This notation is justified by the geometric theory of integration of differential forms, which lies beyond the scope of these notes.

$$(\nabla\phi)_i = \frac{\partial\phi}{\partial x_i}. \tag{1.20}$$

The divergence of this vector field is the scalar field

$$div(\mathbf{grad}\phi) = \nabla\cdot(\nabla\phi) = \nabla^2\phi = \frac{\partial^2\phi}{\partial x_1^2} + \frac{\partial^2\phi}{\partial x_2^2} + \frac{\partial^2\phi}{\partial x_3^2}, \tag{1.21}$$

where we have used various possible notations. In particular, the operator ∇^2 is known as the *Laplacian*.

Naturally, the divergence theorem can be applied to the case of a vector field obtained as the gradient of a scalar field. The result is

$$\int_{\mathcal{D}} \nabla^2\phi\, dV = \int_{\partial\mathcal{D}} \nabla\phi\cdot n\, dA. \tag{1.22}$$

What is the meaning of the term $\nabla\phi\cdot n$? Clearly, this linear combination of partial derivatives is nothing but the *directional derivative* of the function ϕ in the direction of the unit vector $\mathbf{n}$. If we denote this derivative by $d\phi/dn$, we can write the statement of the divergence theorem for the gradient of a scalar field as

$$\int_{\mathcal{D}} \nabla^2\phi\, dV = \int_{\partial\mathcal{D}} \frac{d\phi}{dn}\, dA. \tag{1.23}$$

1.4 Ordinary Differential Equations

1.4.1 Vector Fields as Differential Equations

As we have seen, given a differentiable curve in $\mathbb{R}^n$, one can define the tangent vector at each of its points. The theory of ODEs can be regarded geometrically as providing the answer to the inverse question. Namely, given a vector field in $\mathbb{R}^n$ and a point p_0 in $\mathbb{R}^n$, can one find a curve γ through p_0 whose tangent vector at every point p of γ coincides with the value of the vector field at p?

Let us try to clarify this idea. A *vector field* in $\mathbb{R}^n$is given by a map

$$\mathbf{v} : \mathbb{R}^n \to \mathbb{R}^n, \tag{1.24}$$

or, in components,

$$v_i = v_i(x_1, \ldots, x_n) \qquad i = 1, \ldots, n. \tag{1.25}$$

According to our statement above, we are looking for a curve

$$x_i = x_i(t), \tag{1.26}$$

satisfying the conditions

$$\frac{dx_i(t)}{dt} = v_i(x_1(t), \ldots, x_n(t)) \qquad i = 1, \ldots, n, \tag{1.27}$$

and the initial conditions (of passing through a given point p_0 with coordinates x_i^0)

$$x_i(t_0) = x_i^0, \tag{1.28}$$

for some (initial) value t_0 of the parameter t. We clearly see that the geometric statement of tangency to the given vector field translates into the analytic statement of Eq. (1.27), which is nothing but a system of ordinary differential equations (ODEs) of the first order.

Remark 1.5 If the vector field vanishes at a point P, the solution of (1.27) through P is constant, so that the entire curve collapses to a point. We say that P is an *equilibrium* position of the system.

Using the notation of Eq. (1.5), the system (1.27) can be written as

$$\frac{d\mathbf{r}}{dt} = \mathbf{v}. \tag{1.29}$$

The system is *linear* if it can be written as

$$\frac{d\mathbf{r}}{dt} = \mathbf{A}\,\mathbf{r}, \tag{1.30}$$

where $\mathbf{A}$ is a square constant matrix.

1.4.2 Geometry Versus Analysis

Understanding of mathematical concepts is often enhanced by approaching a problem from various angles, just as we can look at a sculpture from different points of view to better imbibe its meaning. In the realm of mathematics, perhaps controlled by different sides of the brain, we can roughly distinguish between two modes of thought: geometrical and analytical. In the instance at hand, the geometric viewpoint is represented by the notion of vector field, while its analytic counterpart is a system of first-order ordinary differential equations. A solution of this system corresponds to an *integral curve* of the vector field, namely, a curve that is everywhere tangent to the background vector field. We note that we are talking about a parametrized curve, so that we are requiring that the vector tangent to the parametrized integral curve must be exactly equal (and not just proportional) to the vector field. The parameter t emerging from the solution process itself may or may not have an intrinsic physical meaning in a given context. Moreover, it should be clear that this parameter is at most determined up to an additive constant. This arbitrariness can be removed by specifying the value $t_0 = 0$ at the 'initial' point x_i^0.

An important question is whether or not a system of ODEs with given initial conditions always has a solution and, if so, whether the solution is unique. Translating this question into geometrical terms, we ask whether, given a vector field, it is always possible to find a (unique) integral curve through a given point P. Geometrical intuition tells us that, as long as the field is sufficiently regular, we can advance a small step in the direction of the local vector at P to reach a nearby point P' and then repeat the process to a nearby point P'', and so on, to obtain at least a small piece of a curve. This intuition, aided by the power of geometric visualization, turns out to be correct and is formalized in the *existence and uniqueness theorem*, which is briefly discuseed in Sect. 1.4.7.

1.4.3 An Example

For illustrative purposes, let us work in the plane ($n = 2$) and let us propose the following vector field (which will be later related to a very specific physical application; can you guess which?)

$$\begin{cases} v_1 = \quad x_2 \\ v_2 = -\sin x_1 \end{cases} \tag{1.31}$$

This vector field is illustrated in Fig. 1.6.

Let us now arbitrarily choose the initial point p_0 with coordinates $x_1^0 = -9, x_2^0 = 2$. We want to find a curve that, passing through this point, is always tangent to the local value of the given vector field. In other words, we want to solve the following nonlinear system of ODEs

$$\frac{dx_1}{dt} = x_2, \tag{1.32}$$

$$\frac{dx_2}{dt} = -\sin x_1. \tag{1.33}$$

The corresponding Mathematica code and plot are shown in Fig. 1.7.

Let us now consider the same example but with different initial conditions, closer to the origin, such as $x_1^0 = -1.5, x_2^0 = 1$. The corresponding Mathematica code and plot are shown in Fig. 1.8.

```
VectorPlot[{x2, −Sin[x1]}, {x1, −9,9}, {x2, −3,3}, AspectRatio → 0.4, FrameLabel →
{X1, X2}, VectorStyle → Black, PlotRange → {{−10,10}, {−4,4}}]
```

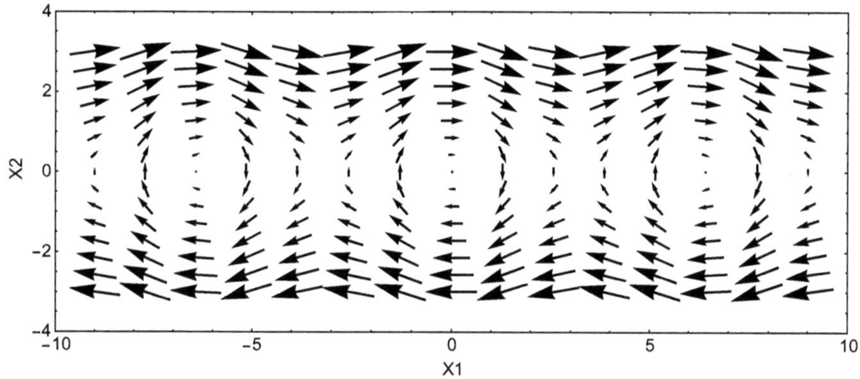

Fig. 1.6 Vector field associated with the system of ODEs (1.31)

```
curve = NDSolve[{x1'[t] == x2[t], x2'[t] == −Sin[x1[t]], x1[0] == −9, x2[0] =
          = 2}, {x1[t], x2[t]}, {t, 0,7.5}]

ParametricPlot[Evaluate[{x1[t], x2[t]}/. curve], {t, 0,7.5}, PlotRange → {{−10,10}, {−4,4}}, PlotStyle
          → Black]
```

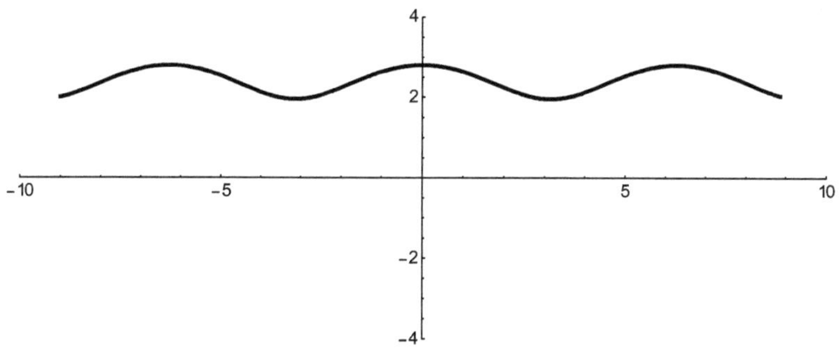

Fig. 1.7 A solution of the system

We obtain a qualitatively different type of solution, represented by a closed curve. To emphasize this fact, we show in Fig. 1.9 a plot of the two integral curves just obtained hovering over the vector field in the background. We can clearly see that the curves are indeed tangential to the vector field at each point!

Remark 1.6 Note that the points with coordinates $(k\pi, 0)$, where k is an integer, are equilibrium positions of the system. The behaviour around an equilibrium point (stability, instability) can be determined by linearizing the system in a small neighbourhood of the point. In our case, the linearized version becomes $d(\Delta x_1)/dt = \Delta x_2$

```
curve1 = NDSolve[{x1'[t] == x2[t], x2'[t] == -Sin[x1[t]], x1[0] == -1.5, x2[0] =
        = 1}, {x1[t], x2[t]}, {t, 0,9}]

ParametricPlot[Evaluate[{x1[t], x2[t]}/. curve1], {t, 0,9}, PlotRange
        → {{-10,10}, {-4,4}}, PlotStyle → Black]
```

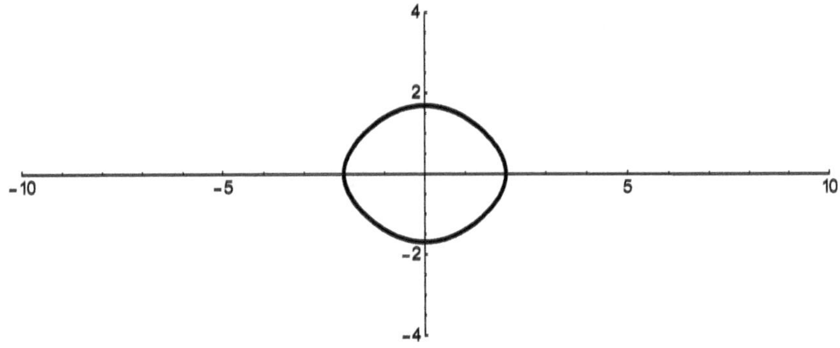

Fig. 1.8 A solution of the system

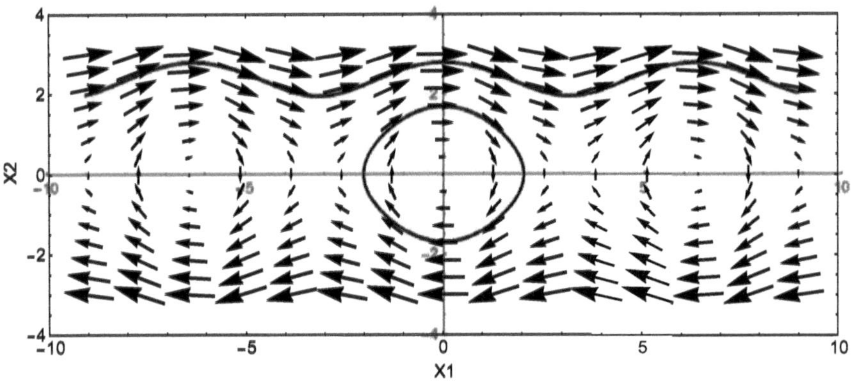

Fig. 1.9 Solutions as integral curves of the vector field

and $d(\Delta x_2)/dt = (-1)^{k+1}\Delta x_1$, where Δx_1, Δx_2 are the incremental variables, so that the linearized system has to be studied in the vicinity of the origin.

1.4.4 *Autonomous and Non-autonomous Systems*

A system of ODEs such as (1.27) is called *autonomous*, a word meant to indicate the fact that the given vector field does not depend on the parameter. A more general, *non-autonomous*, system would have the form

$$\frac{dx_i(t)}{dt} = v_i(t, x_1(t), \ldots, x_n(t)) \qquad i = 1, \ldots, n. \tag{1.34}$$

If, as is often the case, the system of equations is intended to represent the evolution of a *dynamical system* (whether in Mechanics or in Economics, etc.) and if the parameter has the intrinsic meaning of time, the explicit appearance of the time variable in the vector field seems to contradict the principle that the laws of nature do not vary in time. As pointed out by Arnold,[4] however, the process of artificially isolating a system, or a part of a system, from its surroundings for the purpose of a simplified description, may lead to the introduction of time-dependent fields.

An important property of the solutions of autonomous systems of ODEs is the *group property*, also known as the *time-shift property*. It states that if $\mathbf{r} = \mathbf{q}(t)$ is a solution of a system of ODEs corresponding to the vector field $\mathbf{v} = \mathbf{v}(\mathbf{r})$, namely if

$$\frac{d\mathbf{q}(t)}{dt} = \mathbf{v}(\mathbf{q}(t)), \tag{1.35}$$

for all t, then the curve $\mathbf{r} = \mathbf{q}(t + s)$, for any fixed s, is also a solution of the same problem. Moreover, the two integral curves coincide. The proof is straightforward. We start by defining the function $\hat{\mathbf{q}}(t) = \mathbf{q}(t + s)$ and proceed to calculate its derivative at some value $t = \tau$ of the parameter. We obtain

$$\left.\frac{d\hat{\mathbf{q}}}{dt}\right|_{t=\tau} = \left.\frac{d\mathbf{q}(t+s)}{dt}\right|_{t=\tau} = \left.\frac{d\mathbf{q}(t)}{dt}\right|_{t=\tau+s} = \mathbf{v}(\mathbf{q}(\tau + s)) = \mathbf{v}(\hat{\mathbf{q}}(\tau)). \tag{1.36}$$

1.4.5 Higher-Order Equations

In mechanical applications, the use of Newton's law leads in general to systems of second-order (rather than first-order) ODEs. Nevertheless, it is not difficult to show that an ordinary differential equation of order n is equivalent to a system of n first-order ODEs. Indeed, let a differential equation of order n be given by

$$\frac{d^n x(t)}{dt^n} = F\left(t, x(t), \frac{dx(t)}{dt}, \ldots, \frac{d^{n-1}x(t)}{dt^{n-1}}\right), \tag{1.37}$$

where F is a differentiable function in each of its $n + 1$ variables. We define the following new dependent variables

[4] In his beautiful book [1]. Another excellent book by the same author, emphasizing applications to Newtonian and Analytical Mechanics, is [2].

$$x_1 = x \qquad x_2 = \frac{dx}{dt} \qquad x_3 = \frac{d^2x}{dt^2} \quad \dots \quad x_n = \frac{d^{n-1}x}{dt^{n-1}}, \tag{1.38}$$

in terms of which the original differential equation can be written as the first-order system

$$\frac{dx_1}{dt} = x_2, \tag{1.39}$$

$$\frac{dx_2}{dt} = x_3, \tag{1.40}$$

$$\vdots$$

$$\frac{dx_{n-1}}{dt} = x_n, \tag{1.41}$$

$$\frac{dx_n}{dt} = F\left(t, x_1, x_2, \dots, x_n\right). \tag{1.42}$$

Thus a system of second-order equations, such as one obtains in the formulation of problems in dynamics of systems of particles (and rigid bodies), can be reduced to a system of first-order equations with twice as many equations. The unknown quantities become, according to the scheme just described, the positions and the velocities of the particles. In this case, therefore, the space of interest is the so-called *phase space*, which always has an even dimension. If the system is non-autonomous, it is sometimes convenient to introduce the odd-dimensional *extended phase space*, which consists of the Cartesian product of the phase space with the time line $\mathbb{R}$. This terminology is widely used even in non-mechanical applications. The vector field corresponding to an autonomous dynamical system is called its *phase portrait* and its integral curves are called the *phase curves*. A careful analysis of the phase portrait of an autonomous dynamical system can often reveal many qualitative properties of its solutions.

1.4.6 First Integrals and Conserved Quantities

Given a vector field $\mathbf{v}$ in $\mathbb{R}^n$ and the corresponding autonomous system of ODEs

$$\frac{d\mathbf{r}(t)}{dt} = \mathbf{v}(\mathbf{r}(t)), \tag{1.43}$$

a *first integral* is a differentiable function

$$\phi : \mathbb{R}^n \to \mathbb{R} \tag{1.44}$$

that attains a constant value over every solution of the system (1.43). In other words, the function $\phi(x_1, \ldots x_n)$ is constant along every integral curve of the vector field. Clearly, any constant function is, trivially, a first integral. We are, therefore, only interested in non-constant first integrals, which are the exception rather than the rule. Whenever a non-constant first integral exists, it is usually of great physical interest, since it represents a *conserved quantity*. A mechanical system is said to be *conservative* if the external forces can be derived from a scalar *potential* $U : \mathbb{R}^n \to \mathbb{R}$, in which case the total energy of the system (kinetic plus potential) is conserved.

Let a mechanical system with n degrees of freedom (such as a collection of springs and masses) be described by the matrix equation

$$\mathbf{M}\,\ddot{\mathbf{r}} = \mathbf{f}(\mathbf{r}), \tag{1.45}$$

where the constant *mass matrix* $\mathbf{M}$ is symmetric and $\mathbf{f}$ is the vector of external forces. The position vector $\mathbf{r}$ is measured in an inertial frame of reference and superimposed dots indicate time derivatives. In many instances (such as when forces are produced by a gravitational or electrostatic field) the external forces derive from a scalar potential $U = U(\mathbf{r})$ according to the prescription

$$\mathbf{f} = -\frac{\partial U(\mathbf{r})}{\partial \mathbf{r}}, \tag{1.46}$$

or, in components,

$$f_i = -\frac{\partial U}{\partial x_i} \qquad i = 1, \ldots, n. \tag{1.47}$$

If we consider the *kinetic energy* T as the scalar function

$$T = T(\dot{\mathbf{r}}) = \frac{1}{2}\,\dot{\mathbf{r}}^T \mathbf{M}\,\dot{\mathbf{r}}, \tag{1.48}$$

the *total energy* E can be defined as

$$E = T + U. \tag{1.49}$$

Let $\mathbf{r} = \mathbf{r}(t)$ be a solution of the system (1.45) for some initial conditions $\mathbf{r}(0) = \mathbf{r}_0$ and $\dot{\mathbf{r}}(0) = \mathbf{u}_0$. Let us calculate the derivative of the total energy E with respect to t along this solution. Using the chain rule of differentiation, we can write

$$\frac{dE}{dt} = \dot{\mathbf{r}}^T \mathbf{M}\,\ddot{\mathbf{r}} - \dot{\mathbf{r}}^T \mathbf{f}, \tag{1.50}$$

where we have exploited the symmetry of the mass matrix and the potential relation (1.46). Collecting terms and enforcing (1.45) (since we have assumed $\mathbf{r}(t)$ to be a solution of this system), we obtain

$$\frac{dE}{dt} = \dot{\mathbf{r}}^T \left(\mathbf{M}\,\ddot{\mathbf{r}} - \mathbf{f}\right) = 0, \tag{1.51}$$

which proves that E is a constant along every trajectory of the system. The value of this constant is uniquely determined by the initial conditions.

For conservative mechanical systems with a single degree of freedom ($n = 1$), the integral curves in the phase space coincide with the level sets of the total energy, as described in Box 1.1. This remark facilitates the qualitative analysis of such systems. For a fuller treatment, consult Chap. 2 of [1] or Sect. 2.12 of [2], which are useful for the solution of Exercises 1.3 and 1.5.

Box 1.1 Architecture and phase portraits. Roughly speaking, a *shell* is a surface with thickness. In industrial architecture, particularly when using reinforced-concrete, it is not uncommon to find *translational shells*. These structures are generated by translating a plane curve while riding upon another plane curve acting as a guide and lying on a perpendicular plane. Translational surfaces are a particular case of surfaces generated by rigid motions of a curve. They were first defined and studied by the French mathematician Jean-Gaston Darboux (1842–1917). In a Cartesian coordinate system x, y, z, the function

$$z = f(x) + g(y)$$

where f and g are functions of one variable, is a translational surface. Either curve, $f(x)$ or $g(y)$, can be considered as the guide for the translation of the other. After a rainfall, if the surface can contain water, the water level will be bounded by a *level curve* of the surface. On the other hand, it is clear from Eqs. (1.46), (1.48) and (1.49) that the total energy, in the case of a one-degree-of-freedom system, can be represented as a translational surface in the space of coordinates (x, $v = \dot{x}$, E). Since in a conservative system the total energy is a constant of the motion, we conclude that the trajectories are the level curves of the energy surface! In geometrical terms, this remark provides a way to visualize the various types of trajectories for a one-degree-of-freedom system by imagining the parabola representing the kinetic energy traveling over the graph of the potential energy as the guide, and then visualizing the level curves. The graphs below correspond to $M = 8$ and $U = x^2(x^2 - 4)$.

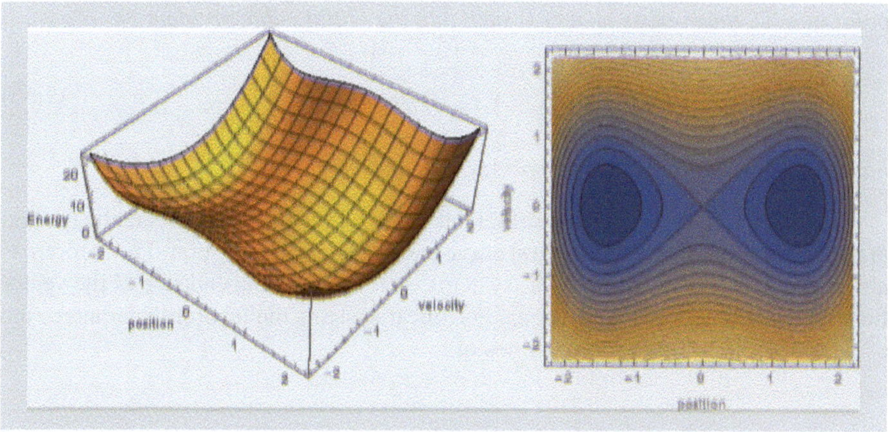

1.4.7 Existence and Uniqueness

When describing, in Sect. 1.4.2, an intuitive geometric way to visualize the construction of an integral curve of a vector field (by moving piecewise along the vectors of the field), we mentioned that the field must be 'sufficiently regular'. It may appear that mere continuity of the field would be sufficient for this intuitive picture to make sense. A more rigorous analysis of the problem, however, reveals that a somewhat stronger condition is needed, namely, *Lipschitz continuity*. In the case of a real function of one real variable

$$f : [a, b] \to \mathbb{R}, \tag{1.52}$$

the function is said to be Lipschitz continuous if there exists a non-negative constant K such that

$$\left| \frac{f(x_2) - f(x_1)}{x_2 - x_1} \right| \leq K, \tag{1.53}$$

for all $x_1 \neq x_2$. An example of a continuous function that is not Lipschitz continuous is the function

$$f(x) = +\sqrt{|x|} \tag{1.54}$$

in the interval $[-1, 1]$.

A differentiable function on a closed interval is automatically Lipschitz continuous. The definition of Lipschitz continuity can be extended to functions of several variables in an obvious way.

Theorem 1.2 (Picard-Lindelöf) *Let* $\mathbf{v}$ *be a vector field defined on a closed domain*[5] $\mathcal{D}$ *of* $\mathbb{R}^n$ *and let the components of* $\mathbf{v}$ *be Lipschitz continuous. Then, for each interior*

[5]This domain may be the whole of $\mathbb{R}^n$.

point $\mathbf{r}_0 \in \mathcal{D}$*, there exists an* $\varepsilon > 0$ *such that the initial value problem*

$$\frac{d\mathbf{r}}{dt} = \mathbf{v} \qquad \mathbf{r}(t_0) = \mathbf{r}_0 \tag{1.55}$$

has a unique solution in the interval $[t_0 - \varepsilon, t_0 + \varepsilon]$.

In geometrical terms, given a sufficiently regular vector field, we can always find at each point a small enough integral curve passing through that point. The theorem is also applicable to non-autonomous systems, as long as the dependence of the vector field on the parameter is continuous. For linear systems the theorem guarantees the existence of the solution for all values of t.

1.4.8 Food for Thought

A vector field on $\mathbb{R}^n$ gives rise to a so-called one-dimensional *distribution* on $\mathbb{R}^n$. Indeed, if we consider the line of action of each vector at each point of $\mathbb{R}^n$, we obtain a field of directions or, equivalently, we have at each point a one-dimensional subspace of $\mathbb{R}^n$. A two-dimensional distribution, accordingly, would consists of attaching at each point a plane (that is, a two-dimensional subspace of $\mathbb{R}^n$). In this spirit, and emboldened by the theorem of existence and uniqueness just presented, we may intuitively foresee that we may be able to construct a small *integral surface* whose tangent plane coincides at each point with the plane in the distribution. This prediction, however, is incorrect, as Exercise 1.7 shows.

Some two-dimensional distributions do admit integral surfaces. These special distributions are called *integrable*. The *theorem of Frobenius* provides a necessary and sufficient conditions for a distribution to be integrable. This integrability condition is known as *involutivity*. A fairly accessible treatment of this subject can be found in [3].

Exercises

Excercise 1.1 Show that the expression (1.16) is preserved upon a change of Cartesian coordinates.

Excercise 1.2 Show that in a system of cylindrical coordinates ξ_1, ξ_2, ξ_3 defined by

$$\begin{aligned} x_1 &= \xi_1 \cos \xi_2 \\ x_2 &= \xi_1 \sin \xi_2 \\ x_3 &= \xi_3, \end{aligned}$$

the divergence of a vector field $\mathbf{v}$ with cylindrical components $\hat{v}_1, \hat{v}_2, \hat{v}_3$ is given by

$$\text{div}\,\mathbf{v} = \frac{1}{\xi_1}\frac{\partial(\xi_1 \hat{v}_1)}{\partial \xi_1} + \frac{1}{\xi_1}\frac{\partial \hat{v}_2}{\partial \xi_2} + \frac{\partial \hat{v}_3}{\partial \xi_3}.$$

Excercise 1.3 (a) Show that the system of ODEs given by Eqs. (1.32) and (1.33) can be used to represent the motion of a pendulum in a vertical plane. (b) Describe qualitatively the behaviour of solutions of the two types discussed above. What kind of solution is represented by closed curves in phase space? (c) By changing the initial conditions one can control which of the two types of behaviour will result. Clearly, there exists a locus of points in phase space corresponding to initial conditions that lie precisely in the boundary between the two types of behaviour. This locus is called a *separatrix*. From considerations of conservation of energy, determine the equation of this separatrix. (d) Plot your separatrix and verify numerically (using, for example, the Mathematica package) that indeed a small perturbation of the initial conditions to one side leads to a different behaviour from that caused by a perturbation to the other side of the separatrix.

Excercise 1.4 Draw (approximately) the phase portrait for a damped pendulum, where the damping force is proportional to the angular velocity. Compare with the results for the undamped pendulum and comment on the nature of the solutions in both cases. Consider various values of the damping coefficient. Is there a critical value? Compare your results qualitatively with the corresponding one for a linear spring with and without damping.

Excercise 1.5 A particle moves along the x axis under the force field

$$F(x) = -1 + 3x^2.$$

Draw and analyze the corresponding phase portrait, with particular attention to the level curves of the total energy (which represent the trajectories of the system in phase space). Do not use a computer package.

Excercise 1.6 Show that for a system of masses subjected only to *central forces* (namely, forces passing through a common fixed point in an inertial frame), the vector of angular momentum of the system with respect to that point is conserved. Recall that the angular momentum is the moment of the linear momentum. For the particular case of a single particle, prove that the trajectories are necessarily plane and derive Kepler's law of areas.

Excercise 1.7 Consider two vector fields, $\mathbf{u}$ and $\mathbf{v}$, in $\mathbb{R}^3$ with components

$$\mathbf{u} = (1, 0, 0) \qquad \mathbf{v} = (0, 1, x_1). \tag{1.56}$$

At each point (x_1, x_2, x_3) of $\mathbb{R}^3$ these two vectors determine a plane. In other words, we have defined a two-dimensional distribution in $\mathbb{R}^3$. Attempt a drawing of this distribution around the origin and explain intuitively why this distribution fails to be involutive. Strengthen your argument by assuming that there exists an integral surface with equation $x_3 = \psi(x_1, x_2)$ and show that imposing the condition that its tangent plane belongs to the distribution (at each point in a vicinity of the origin) leads to a contradiction.

References

1. Arnold VI (1973) Ordinary differential equations. MIT Press, Cambridge
2. Arnold VI (1978) Mathematical methods of classical mechanics. Springer, Berlin
3. Chern SS, Chen WH, Lam KS (2000) Lectures on differential geometry. World Scientific, Singapore
4. Courant R (1949) Differential and integral calculus, vol II. Blackie and Son Ltd, Glasgow
5. Marsden JE, Tromba AJ (1981) Vector calculus, 2nd edn. W. H. Freeman, San Francisco
6. Spivak M (1971) Calculus on manifolds: a modern approach to classical theorems of advanced calculus. Westview Press, Boulder

Chapter 2
Partial Differential Equations in Engineering

Many of the PDEs used in Engineering and Physics are the result of applying physical laws of conservation or balance to systems involving *fields*, that is, quantities defined over a continuous background of two or more dimensions, such as space and time. Under suitable continuity and differentiability conditions, a generic balance law in both global (integral) and local (differential) forms can be derived and applied to various contexts of practical significance, such as Traffic flow, Solid Mechanics, Fluid Mechanics and Heat Conduction.

2.1 Introduction

Partial differential equations arise quite naturally when we apply the laws of nature to systems of continuous extent. We speak then of *field theories*. Thus, whereas the analysis of the vibrations of a chain of masses interconnected by springs gives rise to a system of ODEs, the dynamic analysis of a bar, where the mass is smeared out continuously over the length of the bar, gives rise to a PDE. From this simple example, it would appear that PDEs are a mere generalization of their ordinary counterparts, whereby a few details need to be taken care of. This false impression is exacerbated these days by the fact that numerical procedures, that can be implemented as computer codes with relative ease, do actually approximate the solutions of PDEs by means of discrete systems of algebraic equations. This is clearly a legitimate thing to do, but one must bear in mind that, unless one possesses a basic knowledge of the qualitative aspects of the behaviour of the continuous system, the discrete approximation may not be amenable to a correct interpretation.

One hardly needs to defend the study of PDEs on these grounds, since they stand alone as one of the greatest intellectual achievements of the human race in its attempt to understand the physical world. Need one say more than the fact that from solid and fluid mechanics all the way to quantum mechanics and general relativity, the

M. Epstein, *Partial Differential Equations*, Mathematical Engineering,
DOI 10.1007/978-3-319-55212-5_2

language of nature has so far been transcribed into PDEs? There has been recently a trend to declare the emergence of a "new science", in which the prevalent language will be that of cellular automata and other tools that represent the behaviour of complex systems as the result of simple interactions between a very large (but finite) number of discrete sites of events.[1] These models are particularly powerful in applications where the underlying phenomena are too intricate to capture in any degree of detail by means of PDEs. Such is the case in multi-scale phenomena that appear in many modern applications in a variety of fields (biology, environmental engineering, nanomechanics, and so on). It is too early to predict the demise of Calculus, however. As many times in the past (think of quantum mechanics, chaos, economics), it appears that in one way or another the usefulness of mathematical limits (differentiation, integration) is not entirely dependent on whether or not the actual physical system can "in reality" attain those limits. Calculus and differential equations are here to stay just as trigonometry and Euclidean geometry are not likely to go away.

2.2 What is a Partial Differential Equation?

A partial differential equation for a function u of the independent variables $x_1, x_2, \ldots, x_n$ $(n > 1)$ is a relation of the form

$$F\left(x_i,\ u,\ u_{,i},\ u_{,ij},\ \ldots,\ u_{,\underbrace{ijk\ldots}_{m \text{ indices}}}\right) = 0, \qquad i, j, k, \ldots = 1, \ldots, n, \tag{2.1}$$

where F is a function and where we have introduced the following notation: A subscript preceded by a comma, indicates a partial derivative with respect to the corresponding independent variable. If more than one index follows a comma, it is understood that successive derivatives have been taken. Thus, for instance,

$$u_{,i} = \frac{\partial u}{\partial x_i} \qquad u_{,ijk} = \frac{\partial^3 u}{\partial x_k \partial x_j \partial x_i}. \tag{2.2}$$

By abuse of notation, we have listed in Eq. (2.1) just a generic term for each order of differentiation, understanding that all the values of the indices are to be considered. For example, when we write the argument $u_{,ij}$ we mean the n^2 entries in the square matrix of second partial derivatives.[2] The requirement $n > 1$ is essential, otherwise (if $n = 1$) we would have an ordinary differential equation. The highest order of

[1]This point of view is advocated in [4] with particular force by Stephen Wolfram, a physicist and the creator of the Mathematica code (which, ironically, is one of the best tools in the market for the solution of differential equations).

[2]On the other hand, recalling the equality of mixed partial derivatives (under assumptions that we assume to be fulfilled), the number of independent entries of this matrix is actually only $n(n+1)/2$.

differentiation appearing, which in our case we have indicated by m, is called the *order* of the differential equation.

By a *solution* of the PDE (2.1) we mean a function $u = u(x_i) = u(x_1, \ldots, x_n)$ which, when substituted in Eq. (2.1), satisfies it identically within a given domain of the independent variables $x_1, \ldots x_n$. Clearly, in this version of the theory, the proposed solution must necessarily be differentiable at least m times (otherwise we wouldn't even be able to check that the equation is satisfied). The function F, which actually characterizes the particular differential equation being studied and represents the physical laws at hand, is also subject to conditions of continuity and differentiability which we will not stipulate at this point, but we will assume that as many derivatives of this function exist as we need. A PDE is said to be *linear* if F depends linearly on the unknown function u and all its derivatives. The coefficients of a linear PDE may still depend on the independent variables. If they don't, we have a case of a linear equation with constant coefficients. Even such a simple situation is not amenable to a straightforward treatment, like in the case of ODEs. If the function F is linear in the highest derivatives only (namely, on all the derivatives of order m), the PDE is said to be *quasi-linear*. Otherwise (if it depends non-linearly on at least one of the highest derivatives) the equation is *nonlinear*.

A relatively simple case is that for which the number of independent variables is equal to 2. In this case, a solution can be visualized as a surface in the 3-dimensional space with coordinates x_1, x_2, u. It follows from this intuitive picture that the analog of the notion of integral curve is that of an *integral surface* and perhaps that the analog of the initial conditions at a point is the specification of initial conditions along a whole curve through which the integral surface must pass. More difficult is to visualize at this point what might be the analog of the vector field which, as we know, is associated with a system of ODEs. Leaving this issue aside for the moment, let us remark that just as we have systems of ODEs we can also have systems of PDEs. The question of the equivalence of a single PDE of higher order to a system of PDEs of order 1 is somewhat more delicate than its ODE counterpart.

2.3 Balance Laws

One of the primary sources of PDEs in Engineering and Physics is the stipulation of *conservation laws*. Conservation laws or, more generally, *balance laws*, are the result of a complete accounting of the variation in time of the content of an extensive physical quantity in a certain domain. A simple analogy is the following. Suppose that you are looking at a big farming colony (the domain of interest) and you want to focus attention on the produce (the physical quantity of interest). As time goes on, there is a variation in the quantity of food contained in the domain. At any given instant of time, you want to account for the rate of change of this food content. There are some internal *sources* represented in this case by the rate at which the land yields new produce (so and so many tons per week, say). There are also *sinks* (or negative sources) represented by the internal consumption of food by workers and cattle, damage caused by hail and pests, etcetera. We will call these sources and sinks the *production* of the quantity in question. It is measured in units of the original quantity divided by the unit of time. In addition to these internal factors, there is also another type of factors that can cause a change in content. We are referring to exchanges of food through the boundary of the colony. These include the buying and selling of produce that takes place at the gates, the perhaps illegal activities of some members or visitors that personally take some food away to other destinations, etcetera. At any given instant of time, we can estimate the rate at which these exchanges take place at the boundary. We will call these transactions the *flux* of the quantity in question. We may have a flux arising also from the fact that the boundary of the domain of interest is changing (encroached by an enemy or by natural causes, etcetera). Assuming that we have accounted for every one of these causes and that we believe in the principles of causality and determinism (at least as far as the material world is concerned), we may write the generic equation of balance as

$$\frac{d\ \text{content}}{dt} = \text{production} + \text{flux}, \tag{2.3}$$

where t is the time variable.

In physically meaningful examples (balance of energy, momentum, mass, electric charge, and so on), it is often the case that the content, the production and the flux are somehow distributed (smeared) over the volume (in the case of the content and the production) or over the area of the boundary (in the case of the flux). In other words, these magnitudes are given in terms of *densities*, which vary (continuously, say) from point to point and from one instant to the next. It is precisely this property (whether real or assumed) that is responsible for the fact that we can express the basic equation of balance (2.3) in terms of differential equations. Indeed, the differential equations are obtained by assuming that Eq. (2.3) applies to any sub-domain, no matter how small.

2.3.1 The Generic Balance Equation

Let U represent an extensive quantity for which we want to write the equation of balance. We assume this quantity to be scalar, such as mass, charge or energy content.[3] Consider a spatial region ω fixed in $\mathbb{R}^3$ and representing a subset of the region of interest. Our four independent variables are the natural coordinates x_1, x_2, x_3 of $\mathbb{R}^3$ and the time variable t.[4] When we say that U is an *extensive* quantity, we mean that we can assign a value of U (the *content*) to each such subset ω. On physical grounds we further assume that this set function is *additive*. By this we understand

[3]Vector quantities, such as linear and angular momentum, can be treated in a similar way by identifying U alternatively with each of the components in a global Cartesian frame of reference.

[4]Consequently, we will not strictly adhere to the notational convention (2.2).

that the total content in two disjoint subsets is equal to the sum of the contents in each separate subset. Under suitable continuity conditions, it can be shown that the content of an extensive quantity U is given by a density $u = u(x_1, x_2, x_3, t)$ in terms of an integral, namely,

$$U = \int_{\omega} u \, d\omega. \tag{2.4}$$

It is clear that this expression satisfies the additivity condition. The units of u are the units of U divided by a unit of volume.

Similarly, the *production* P is assumed to be an extensive quantity and to be expressible in terms of a production density $p = p(x_1, x_2, x_3, t)$ as

$$P = \int_{\omega} p \, d\omega. \tag{2.5}$$

The units of p are the units of U divided by a unit of volume and by the time unit. We adopt the sign convention that a positive p corresponds to creation (source) and a negative p corresponds to annihilation (sink).

The *flux* $\mathcal{F}$ represents the change in content per unit time flowing through the boundary $\partial\omega$, separating the chosen subset ω from the rest of the region of interest. In other words, the flux represents the contact interaction between adjacent subsets. A remarkable theorem of Cauchy shows that under reasonable assumptions the flux is governed by a vector field, known as the *flux vector* $\mathbf{f}$. More precisely, the inflow per unit area and per unit time is given by

$$d\mathcal{F} = (\mathbf{f} \cdot \mathbf{n}) da, \tag{2.6}$$

where $\mathbf{n}$ is the exterior unit normal to da. The significance of this result can be summarized as follows:

1. The 'principle of action and reaction' is automatically satisfied. Indeed at any given point an element of area da can be considered with either of two possible orientations, corresponding to opposite signs of the unit normal $\mathbf{n}$. Physically, these two opposite vectors represent the exterior unit normals of the sub-bodies on either side of da. Thus, what comes out from one side must necessarily flow into the other.
2. All boundaries $\partial\omega$ that happen to have the same common tangent plane at one point transmit, at that point, exactly the same amount of flux. Higher order properties, such as the curvature, play no role whatsoever in this regard. In fact, this is the main postulate needed to prove Cauchy's theorem.
3. The fact that the amount of flux depends *linearly* on the normal vector (via the dot product) conveys the intuitive idea that the intensity and the angle of incidence of the flowing quantity are all that matter. If you are sun-tanning horizontally at

high noon in the same position as two hours later, you certainly are taking in more radiation per unit area of skin in the first case.

We are now in a position of implementing all our hypotheses and conclusions into the basic balance Eq. (2.3). The result is

$$\frac{d}{dt}\int_{\omega} u\, d\omega = \int_{\omega} p d\omega + \int_{\partial\omega} \mathbf{f}\cdot\mathbf{n}\, da. \tag{2.7}$$

This equation represents the *global balance equation* for the volume ω. It should be clear that this equation is valid under relatively mild conditions imposed on the functions involved. Indeed, we only need the density u to be differentiable with respect to time and otherwise we only require that the functions be integrable. This remark will be of great physical significance when we study the propagation of shocks. In the case of a content u and a flux vector $\mathbf{f}$ which are also space-wise differentiable, we can obtain a *local* version of the generic balance equation. This local ('infinitesimal') version is a partial differential equation. To derive it, we start by observing that, due to the fact that the volume ω is fixed (that is, independent of time), the order of differentiation and integration on the left-hand side of Eq. (2.7) can be reversed, that is,

$$\frac{d}{dt}\int_{\omega} u\, d\omega = \int_{\omega} \frac{\partial u}{\partial t} d\omega. \tag{2.8}$$

Moreover, the surface integral on the right-hand side of Eq. (2.7) is the flux of a vector field on the boundary of a domain and is, therefore, amenable to be treated by means of the divergence theorem according to Eq. (1.18), namely,

$$\int_{\partial\omega} \mathbf{f}\cdot\mathbf{n}\, da = \int_{\omega} \mathrm{div}\mathbf{f} d\omega. \tag{2.9}$$

Collecting all the terms under a single integral we obtain the global balance equation in the form

$$\int_{\omega} \left(\frac{\partial u}{\partial t} - p - \mathrm{div}\mathbf{f}\right) d\omega = 0. \tag{2.10}$$

This equation is satisfied identically for any arbitrary sub-domain ω. If the integrand is continuous, however, it must vanish identically. For suppose that the integrand is, say, positive at one point within the domain of integration. By continuity, it will also be positive on a small ball B around that point. Applying the identity (2.10) to this sub-domain B, we arrive at a contradiction. We conclude, therefore, that a necessary and sufficient condition for the global balance equation to be satisfied identically for arbitrary sub-domains is the identical satisfaction of the partial differential equation

$$\frac{\partial u}{\partial t} - p - \text{div}\mathbf{f} = 0. \tag{2.11}$$

This is the generic equation of balance in its local (differential) form. It is a single PDE for a function of 4 variables, x_1, x_2, x_3 and $x_4 = t$.

2.3.2 The Case of Only One Spatial Dimension

There are several reasons to present an independent derivation of the generic law of balance for the case of a single spatial dimension. The first reason is that in the case of just one dominant spatial dimension (waves or heat flow in a long bar, current in a wire, diffusion of pollutants in a tube, etcetera), the divergence theorem mercifully reduces to the statement of the fundamental theorem of calculus of one variable (roughly speaking: "differentiation is the inverse of integration"). Notice that we still are left with two independent variables, one for the spatial domain (x) and one for the time dependence (t). Another important reason has to do with the peculiar nature of a domain in $\mathbb{R}$ as compared with domains in higher dimensions. If the spatial domain is two-dimensional, such as a membrane, its boundary is the perimeter curve, while the upper and lower faces of the membrane are identified with the interior points. For a three-dimensional domain, the boundary is the whole bounding surface. On the other hand, a closed connected domain in $\mathbb{R}$ is just a closed interval $[a, b]$, with $a < b$. Its boundary consists of just two distinct points, as shown in Fig. 2.1. Moreover, the exterior normal to the boundary is defined at those points only, as a unit vector at a pointing in the negative direction and a unit vector at b pointing in the positive direction of the real line. The flux vector $\mathbf{f}$ and the velocity vector $\mathbf{v}$ have each just one component and can be treated as scalars. Physically, we may think of U as the content of some extensive quantity in a wire or a long cylindrical bar. It is important to realize that the lateral surface of this wire does not exist, in the sense that it is not part of the boundary. On the contrary, the points on this

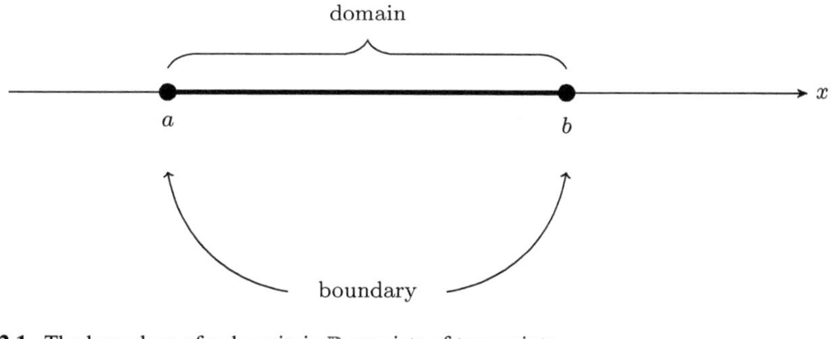

Fig. 2.1 The boundary of a domain in $\mathbb{R}$ consists of two points

(vanishingly small) lateral surface are identified precisely with the interior points of the wire.

If we assume that the quantity $U = U(t)$ is continuously distributed throughout the domain, we can express it in terms of a density $u = u(x, t)$ per unit length of the bar as

$$U = \int_a^b u \, dx. \tag{2.12}$$

Similarly, the production $P = P(t)$ can be expressed in terms of a density $p = p(x, t)$ per unit length and per unit time as

$$P = \int_a^b p \, dx. \tag{2.13}$$

As a sign convention, we assume that a positive p corresponds to creation (source) and a negative sign to annihilation (sink).

The flux term requires some further discussion. Clearly, if we cut the bar into two pieces and we focus attention on one of these pieces alone, as far as the quantities U and P are concerned, all we have to do is change the limits of integration in Eqs. (2.12) and (2.13). On the other hand, by the process of cutting, we have created a new boundary point at each of the sub-bodies and a corresponding flux. If we assume (following an idea similar to Newton's law of action and reaction) that whatever flow enters through the new boundary into one of the parts must necessarily be coming out of the new boundary of the other part, we realize that the flux is best represented by the dot product of a *flux vector* $\mathbf{f} = \mathbf{f}(x, t)$ with the unit vector $\mathbf{n}$ at the boundary *pointing in the outward direction of the part under study*. This situation is illustrated in Fig. 2.2.

If the flux vector points to the right, the flux through the cross section, as shown in the figure, will be positive (an inflow) for the left part of the bar and negative

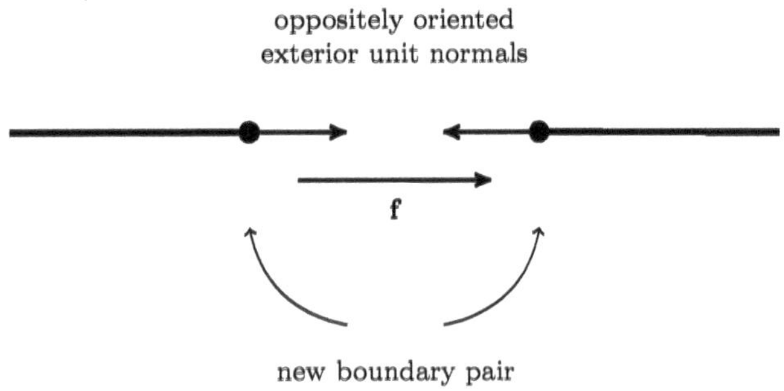

Fig. 2.2 The flux vector

(and of equal absolute value) for the right part of the bar. Notice that in this simple one-dimensional case, the flux vector has only one component, which we denote by $f = f(x,t)$. Nevertheless, the concept of flux vector is very important and can be used directly in two- and three- dimensional spatial contexts. For the actual boundary of the body, the flux vector may be specified as a *boundary condition*, depending on the specific problem being solved.

Introducing our specific expressions for content, production and flux in the generic balance equation (2.3), we obtain

$$\frac{d}{dt}\int_a^b u(x,t)\,dx = \int_a^b p(x,t)\,dx + f(b,t) - f(a,t). \tag{2.14}$$

As far as sign convention for the flux is concerned, we have assumed that a positive scalar flux is inwards, into the body. What this means is that if the flux vector points outwards, the scalar flux is actually inwards. If this convention is found unnatural, all one has to do is reverse the sign of the last two terms in Eq. (2.14).[5]

This is the equation of balance in its *global form*. It is not yet a partial differential equation. It is at this point that, if we wish to make the passage to the local form of the equation, we need to invoke the fundamental theorem of calculus (or the divergence theorem in higher dimensional contexts). Indeed, we can write

$$f(b,t) - f(a,t) = \int_a^b \frac{\partial f(x,t)}{\partial x}\,dx. \tag{2.15}$$

We obtain, therefore,

$$\frac{d}{dt}\int_a^b u(x,t)\,dx = \int_a^b p(x,t)\,dx + \int_a^b \frac{\partial f}{\partial x}\,dx. \tag{2.16}$$

If we consider that the integral limits a and b, though arbitrary, are independent of time, we can exchange in the first term of the equation the derivative with the integral, namely,

$$\int_a^b \frac{\partial u}{\partial t}\,dx = \int_a^b p(x,t)\,dx + \int_a^b \frac{\partial f}{\partial x}\,dx, \tag{2.17}$$

identically for all possible integration limits. We claim now that this identity is possible only if the integrands themselves are balanced, namely, if

[5]This common policy is adopted in [3]. This is an excellent introductory text, which is highly recommended for its clarity and wealth of examples.

$$\frac{\partial u}{\partial t} = p(x,t) + \frac{\partial f}{\partial x}. \tag{2.18}$$

The truth of this claim can be verified by collecting all the integrands in Eq. (2.17) under a single integral and then arriving at a combined integrand whose integral must vanish no matter what limits of integration are used. Clearly, if the integrand is continuous and does not vanish at some point in the domain of integration, it will also not vanish at any point in a small interval containing that point (by continuity). It will, therefore, be either strictly positive or strictly negative therein. Choosing, then, that small interval as a new domain of integration, we would arrive at the conclusion that the integral does not vanish, which contradicts the assumption that the integration must vanish *for all values* of the limits. We conclude that Eq. (2.18) must hold true.

2.3.3 The Need for Constitutive Laws

When we look at the local form of the balance equation, (2.18), we realize that we have a single equation containing partial derivatives of two unknown functions, u and f. What this is telling us from the physical point of view is that the equations of balance are in general not sufficient to solve a physical problem. What is missing? If we think of the problem of heat transfer through a wire (which is an instance of the law of balance of energy), we realize that the *material properties* have played no role whatsoever in the formulation of the equation of balance. In other words, at some point we must be able to distinguish (in these macroscopic phenomenological models in which matter is considered as a continuum) between different materials. Copper is a better heat conductor than wood, but the law of balance of energy is the same for both materials! The missing element, namely the element representing the response of a specific medium, must be supplied by means of an extra equation (or equations) called the *constitutive law* of the medium. Moduli of elasticity, heat conductivities, piezoelectric and viscosity constants are examples of the type of information that may be encompassed by a constitutive law. And what is that the constitutive equation can stipulate? Certainly not the production, since this is a matter of sources and sinks, which can be controlled in principle regardless of the material at hand. Instead, it is the flux vector within the body that will differ from material to material according to the present state of the system. The state of a system is given in terms of some local variables of state $s_1, s_2, \ldots, s_k$ (positions, temperatures, velocity gradients, and so on), so that both the flux f and the content density u may depend on them. The constitutive law is then expressed by equations such as

$$u = \hat{u}\,(s_1(x,t), \ldots, s_k(x,t), x) \tag{2.19}$$

and

$$f = \hat{f}\,(s_1(x,t), \ldots, s_k(x,t), x)\,. \tag{2.20}$$

The reason that we have included a possible explicit dependence on x is that the properties of the medium may change from point to point (as is the case in the so-called functionally graded bodies, for instance). In principle, these properties could also change in time, as is the case in processes of aging (biological or otherwise). In some cases, a single variable of state is enough to characterize the system, so that ultimately Eq. (2.18) becomes a PDE for the determination of this variable of state as a function of space and time. Sometimes, it is possible to adopt the density u itself as a single variable of state, so that the constitutive law simply reads

$$f = \hat{f}(u(x,t), x). \tag{2.21}$$

In this case, substituting the constitutive law into (2.18), we obtain

$$\frac{\partial u}{\partial t} = p + \frac{\partial \hat{f}}{\partial u}\frac{\partial u}{\partial x} + \frac{\partial \hat{f}}{\partial x}. \tag{2.22}$$

It is often convenient to adopt a subscript notation for partial derivatives of the unknown field variable $u = y(x,t)$, such as

$$u_x = \frac{\partial u}{\partial x} \qquad u_t = \frac{\partial u}{\partial t} \qquad u_{xx} = \frac{\partial^2 u}{\partial x^2} \qquad u_{xt} = \frac{\partial^2 u}{\partial t \partial x} \quad \ldots \tag{2.23}$$

Notice that, since there is no room for confusion, we don't place a comma before the subscripts indicating derivatives, as we did in (2.2). In this compact notation, Eq. (2.22) reads

$$u_t - \frac{\partial \hat{f}}{\partial u}\, u_x = p + \frac{\partial \hat{f}}{\partial x}. \tag{2.24}$$

We have purposely left the partial derivatives of the constitutive function $\hat{f}$ unaffected by the subscript notation. The reason for this is that the constitutive function $\hat{f}$ is not an unknown of the problem. On the contrary, it is supposed to be known as that part of the problem statement that identifies the material response. Its partial derivatives are also known as some specific functions of u and x. Notice that in the case of a homogeneous material, the last term in Eq. (2.24) vanishes.

In the terminology introduced in the previous section, Eq. (2.24) is a first order, quasi-linear PDE. If the constitutive function $\hat{f}$ happens to be a linear function of u, the PDE becomes linear. The linearity of constitutive laws is still one of the most common assumptions in many branches of engineering (for example: Hooke's law, Ohm's law, Fourier's law, Darcy's law, etcetera, are not actual laws of nature but constitutive assumptions that are useful linear approximations to the behaviour of some materials within certain ranges of operation). Notice that in the examples just mentioned, the constitutive laws are expressed in terms of space derivatives of state variables (respectively, displacement, electric potential, temperature and pressure). As a result, the equation of balance combined with the constitutive law yields a second order PDE. The theory of a single first-order PDE is comparable in its precision and implementation to the theory of systems of ODEs. This is not the case for higher order PDEs or for systems of first order PDEs, as we shall see later. At this point, however, we are only interested in illustrating the emergence of PDEs of any order and type from well-defined engineering contexts, without much regard for their possible solutions. Accordingly, in the next section, we will display several instances of balance laws, which constitute a good (but by no means the only) source of PDEs in applications.

2.4 Examples of PDEs in Engineering

2.4.1 *Traffic Flow*

A comprehensive review of models for traffic flow is beyond our present scope. Instead, we present here a simplified version of the fundamental equation, based on the assumptions that the road is of a single lane and that (within the portion of road being analyzed) there are no entrances or exits. The quantity we want to balance is the content of cars. We, therefore, interpret $u = u(x, t)$ as the car density at the point x along the road at time t. Since we have assumed no entrances or exits, the production term p vanishes identically. The flux term f has the following physical interpretation: At any given cross section of the road and at a given instant of time, it measures the number of cars per unit time that pass through that cross section or, more precisely, the number of cars per unit time that enter one of the portions of road to the right or left of the cross section. With our (counter-intuitive) sign convention, a positive value of f corresponds to an inflow of cars. We have seen that the flux is actually governed by a flux vector $\mathbf{f}$. Denoting by $\mathbf{v} = \mathbf{v}(x, t)$ the car-velocity field, we can write

$$\mathbf{f} = -u\ \mathbf{v}. \tag{2.25}$$

In other words, if the velocity points in the direction of the exterior normal to the boundary (so that the dot product is positive) the term $u\ \mathbf{v}$ measures the number of cars that in a unit of time are coming out through that boundary. Since in our case everything is one-dimensional, the velocity vector is completely defined by its component v along the axis of the road, so that we can write

$$f = -u\ v. \tag{2.26}$$

The local balance equation (2.24) for this traffic flow problem reads, therefore,

$$u_t + \frac{\partial(u\ v)}{\partial u}\ u_x = 0. \tag{2.27}$$

The time has come now to adopt some constitutive law. Clearly, the velocity of the cars may depend on a large number of factors, including the time of day, the weather, the traffic density, etcetera. In the simplest model, the velocity will depend only on the traffic density, with larger densities giving rise to smaller speeds. From practical considerations, since cars have a finite length, there will be an upper bound u_{max} for the density, and it is sensible to assume that when this maximum is attained the traffic comes to a stop. On the other hand, we may or may not wish to consider an upper limit v_{max} for the speed, when the traffic density tends to zero. If we do, a possible constitutive equation that we may adopt is

$$v = v_{max}\left(1 - \frac{u}{u_{max}}\right). \tag{2.28}$$

If, on the other hand, we do not want to impose a speed limit in our model, a possible alternative constitutive law is

$$v = k \ln \frac{u_{max}}{u}, \tag{2.29}$$

where k is a positive constant (which perhaps varies from road to road, as it may take into consideration the quality of the surface, the width of the lane, and so on).

Introducing the constitutive law (2.28) into our balance law (2.27), we obtain the quasi-linear first-order PDE

$$u_t + \left(1 - 2\frac{u}{u_{max}}\right) v_{max}\, u_x = 0. \tag{2.30}$$

In the extreme case when the speed is independent of the density and equal to a constant, we obtain the *advection equation*

$$u_t + v_{max}\, u_x = 0. \tag{2.31}$$

2.4.2 *Diffusion*

Diffusive processes are prevalent in everyday life. They occur, for example, whenever a liquid or gaseous substance spreads within another (sneezing, pouring milk into a cup of coffee, industrial pollution, etc.). The process of heat flow through a substance subjected to a temperature gradient is also a diffusive process. All these processes are characterized by thermodynamic irreversibility (the drop of milk poured into the coffee will never collect again into a drop).

Consider a tube filled with water at rest in which another substance (the pollutant) is present with a variable concentration $u = u(x, t)$. Let $p = p(x, t)$ be the production of pollutant per unit length and per unit time. This production can be the result of industrial exhaust into the tube, coming from its lateral surface at various points, or of a similar process of partial clean-up of the tube. If there is any influx through the ends of the tube, it will have to be considered as part of the boundary conditions (which we have not discussed yet), rather than of the production term. The flux, just as in the case of traffic flow, represents the amount of pollutant traversing a given cross section per unit time. In the case of traffic flow, we introduced as a variable of state the speed of the traffic, which we eventually related to the car density by means of a constitutive law. In the case of diffusion of a pollutant, on the other hand, it is possible to formulate a sensible, experimentally based, constitutive law directly in terms of the pollutant concentration. The most commonly used law, called Fick's law, states that the flux vector is proportional to the gradient of the concentration, namely,

$$f = D\, u_x, \tag{2.32}$$

where the constant D is the *diffusivity* of the pollutant in water. A moment's reflection reveals that, with our sign convention, if we want the pollutant to flow in the direction of smaller concentrations, the diffusivity must be positive. Introducing these results into the general balance equation (2.18), we obtain

$$u_t - D\, u_{xx} = p. \tag{2.33}$$

This second-order linear PDE is known as the (non-homogeneous)[6] diffusion equation. It is also known as the one-dimensional *heat equation*, in which case u stands for the temperature and the constant D is a combination of the heat capacity and the conductivity of the material.

2.4.3 Longitudinal Waves in an Elastic Bar

Assuming that the particles in a thin cylindrical bar are constrained to move in the axial direction, the law of balance of momentum (Newton's second law) can be seen as a scalar equation. The momentum density (momentum per unit length) is given by $\rho\, A\, v$, ρ being the mass density, A the cross-section area and v the component of the velocity vector. The production term in this case consists of any applied force per unit length (such as the weight, if the bar is held vertically). We will assume for now that there are no applied external forces, so that the production term vanishes identically. The flux associated with the momentum is what we call the stress tensor, which in this case can be represented by a single component σ (perpendicular to the normal cross sections). The balance of momentum[7] reads

[6]The adjective *non-homogeneous*, in this case, refers to the fact that there are sources or sinks, that is, p does not vanish identically. *Material inhomogeneity*, on the other hand, would be reflected in a variation of the value of the diffusivity D throughout the tube.

[7]Neglecting convective terms.

$$\frac{\partial(\rho\, A\, v)}{\partial t} = \frac{\partial(\sigma\, A)}{\partial x}. \tag{2.34}$$

Assuming constant cross section and density, we obtain

$$\rho\, v_t = \sigma_x. \tag{2.35}$$

This balance equation needs to be supplemented with a constitutive law. For a linearly elastic material, the stress is proportional to the strain (ε), that is,

$$\sigma = E\, \varepsilon, \tag{2.36}$$

where E is Young's modulus. Adopting the (axial) displacement $u = u(x, t)$ as a state variable, we have

$$v = u_t, \tag{2.37}$$

by definition of velocity, and

$$\varepsilon = u_x, \tag{2.38}$$

by the kinematic relations of the infinitesimal theory of strain. Putting all these results together we obtain the second-order linear PDE

$$u_{tt} = c^2\, u_{xx}, \tag{2.39}$$

where the constant c is given by

$$c = \sqrt{\frac{E}{\rho}}. \tag{2.40}$$

Equation (2.39) is known as the *one-dimensional wave equation*. The constant c will be later interpreted as the speed of propagation of waves in the medium. A similar equation can be derived for the problem of small transverse vibrations of a string (such as that of a guitar) under tension. In this case, the constant c is given by the square root of the ratio between the tension in the string and its mass per unit length.[8]

2.4.4 Solitons

In many phenomena not governed by the wave equation it is possible to observe the propagation of highly concentrated pulses traveling undistorted at a constant speed. These traveling waves are known as *solitons*. They were first observed in shallow water, where gravity drives the propagation. The equation governing this phenomenon was first derived by Korteweg and de Vries in 1895 and is now known

---

[8]See Sect. 8.1.

as the KdV equation. We will not present its derivation. It reads

$$u_t + u\, u_x + u_{xxx} = 0. \tag{2.41}$$

Here, $u = u(x, t)$ represents a measure of the height of the water in a long channel of constant cross section. The KdV equation is a third-order quasi-linear PDE. It can be brought to the form of a conservation law (2.18) by setting $p = 0$ and

$$f = -\left(\frac{1}{2}u^2 + u_{xx}\right). \tag{2.42}$$

2.4.5 *Time-Independent Phenomena*

If in the examples just presented we eliminate the dependence of the variables on time, namely if $u = u(x)$, we obtain in each case an ODE representing a configuration of steady state or equilibrium of the system. On the other hand, if we were to extend the spatial domain from one to two dimensions (rather than just one, as we have been doing so far), the steady-state equation would still be a PDE in two independent variables. As an example, we will consider the equilibrium configuration of a membrane which has been initially stretched by applying a high uniform tension T (per unit length, say) in all directions and then attached to a rigid plane frame along its perimeter. The membrane thus prepared is then subjected to a transverse load (perpendicular to the plane of the frame) of magnitude $q(x, y)$, where x, y is a system of Cartesian coordinates in the plane of the unloaded membrane. We are interested in calculating the transverse deflection $w = w(x, y)$ corresponding to an equilibrium configuration.

We have already remarked that the transverse vibrations of a tensed string are given by the wave equation (2.39). A careful derivation of the analogous two-dimensional membrane counterpart, as depicted in Fig. 2.3, would lead to the dynamical equation

$$w_{xx} + w_{yy} = -\frac{q}{T} + \frac{\rho h}{T}\, w_{tt}, \tag{2.43}$$

where h is the thickness of the membrane. In the absence of the external loading term q, this is the *two-dimensional wave equation*. If, on the other hand, we seek an equilibrium position under the action of a time-independent load, we obtain the second-order linear PDE

$$w_{xx} + w_{yy} = -\frac{q}{T}. \tag{2.44}$$

This is the *Poisson equation*. If the right-hand side vanishes (no load, but perhaps a slightly non-planar frame) we obtain the *Laplace equation*. These equations appear in many other engineering applications, including fluid mechanics, acoustics, electrostatics and gravitation.

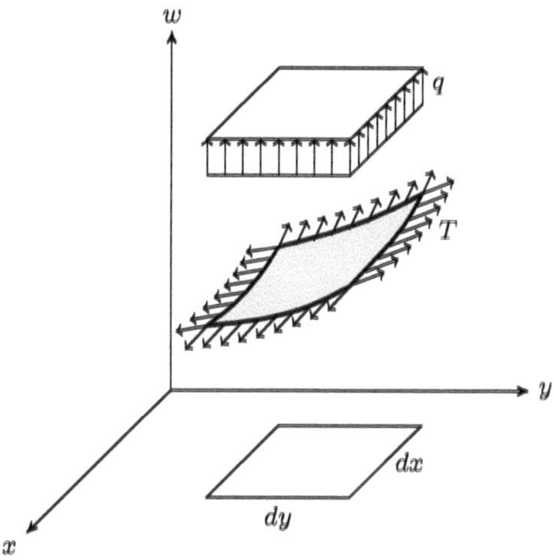

Fig. 2.3 Balance of forces in a membrane

2.4.6 Continuum Mechanics

In Continuum Mechanics [1] the field variables are always associated with a continuous *material body* as the carrier of contents, sources and fluxes. The material body, made up of *material points*, manifests itself through its *configurations* in the physical space $\mathbb{R}^3$. In this brief presentation, we will adhere strictly to the *Eulerian formulation*, which adopts as its theatre of operations the current configuration of the body in space. The domains ω used in the formulation of the generic balance equation must, accordingly, be subsets of the current configuration. In other words, they must be made of spatial points occupied at the current time by material particles. The generic equation of balance in its global form (2.7) or in its local form (2.11), is still applicable. In Continuum Mechanics, however, it is convenient to identify two distinct parts of the total flux $\mathcal{F}$ through the boundary $\partial\omega$ which we call the *convected flux* $\mathcal{F}_c$ and the *physical flux* $\mathcal{F}_p$, that is,

$$\mathcal{F} = \mathcal{F}_c + \mathcal{F}_p. \tag{2.45}$$

The convected flux appears as a natural consequence of having adopted a *fixed spatial volume* ω for the analysis. Since the material particles are moving with a velocity field $\mathbf{v} = \mathbf{v}(x_1, x_2, x_3, t)$, they are in general entering into or exiting from ω. In so doing, they import or export a certain amount of content per unit time. Figure 2.4 makes it clear what this amount $d\mathcal{F}_c$ is at any given elemental area da lying on the boundary of ω. We have

$$d\mathcal{F}_c = -u(\mathbf{v} \cdot \mathbf{n})da, \tag{2.46}$$

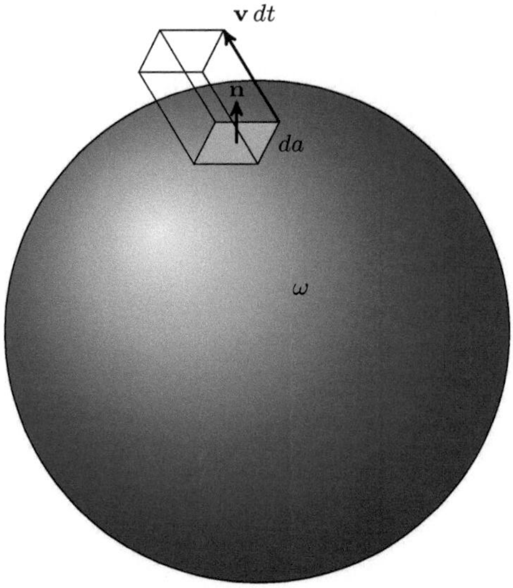

Fig. 2.4 The amount of content flowing out of the volume ω through the area element da during a time dt is proportional to the volume of the parallelepiped shown

where $\mathbf{n}$ is the *exterior unit normal* to da as part of $\partial\omega$. The negative sign indicates that we have assumed an influx as positive. Naturally, if the particle velocity happens to be tangential to the boundary at a point, the convected flux at that point vanishes.

Remark 2.1 Had we assumed a *material volume* as the point of departure (that is, a volume that follows the particles in their motion in space), the corresponding convected flux would have automatically vanished. This simplification would have resulted, however, in the need to use a compensatory *transport theorem* for the calculation of the time variation of an integral on a moving domain. The convected flux is, literally, in the eyes of the beholder.

The second part of the flux, $\mathcal{F}_p$ has the clear physical meaning of the flow of content through the boundary due to causes other than mere motion or rest of the control volume. Thus, for instance, if the content in question is the internal energy of a rigid body at rest, the flux through the boundary represents the conductive heat flux. It is important to notice once more that the physical flux takes place at each internal boundary (not just the external boundary of the body) separating a sub-body from the rest. Cauchy's theorem implies that the physical flux is governed by a flux vector field $\mathbf{f}_p$, as before. Thus, we obtain the global form

$$\frac{d}{dt}\int_{\omega} u\, d\omega = \int_{\omega} p d\omega + \int_{\partial\omega} (-u\mathbf{v} + \mathbf{f}_p)\cdot\mathbf{n}\, da. \tag{2.47}$$

Applying the divergence theorem and following our previous localization argument we obtain the generic law of balance of Continuum Mechanics in its local form as

$$\frac{\partial u}{\partial t} - p - \text{div}(-u\mathbf{v} + \mathbf{f}_p) = 0. \tag{2.48}$$

In terms of the *material derivative*, discussed in Box 2.1, this equation can also be written as

$$\frac{Du}{Dt} - p + u\ \text{div}\mathbf{v} - \text{div}\mathbf{f}_p = 0. \tag{2.49}$$

Box 2.1 The material derivative

The partial time-derivative $\frac{\partial u}{\partial t}$ appearing in Eq. (2.48) describes the rate of change of the density u *at a fixed spatial position* (x_1, x_2, x_3). If we imagine an observer sitting at that position and recording events as time goes on, this partial derivative is the slope of the graph of u versus time as produced by that observer. Thus, if the regime happens to be *steady*, this slope will vanish identically. If we imagine, instead, a hypothetical observer riding with *a fixed material particle*, this second observer will record a different result! The slope of the graph recorded by this moving observer describes the rate of change of u at that particle. This is called the *material derivative* of u, denoted by Du/Dt. In the case of a steady state which is spatially non-constant, clearly the material derivative will not vanish in general. What is the relation between the partial derivative and the material derivative? The particle that at time t passes through the spatial position (x_1, x_2, x_3) will occupy at time $t + dt$ the position $(x_1 + v_1 dt, x_2 + v_2 dt, x_3 + v_3 dt)$, where (v_1, v_2, v_3) are the local components of the velocity. At this point the value of u has, to a first degree of approximation, the value

$$\begin{aligned} &u(x_1 + v_1 dt, x_2 + v_2 dt, x_3 + v_3 dt, t + dt) \\ =&u(x_1, x_2, x_3, t) + \frac{\partial u}{\partial x_1} v_1 dt + \frac{\partial u}{\partial x_2} v_2 dt + \frac{\partial u}{\partial x_3} v_3 dt + \frac{\partial u}{\partial t} dt. \end{aligned}$$

Accordingly, we obtain

$$\begin{aligned} \frac{Du}{Dt} &= \lim_{dt \to 0} \frac{u(x_1 + v_1 dt, x_2 + v_2 dt, x_3 + v_3 dt, t + dt) - u(x_1, x_2, x_3, t)}{dt} \\ &= \frac{\partial u}{\partial t} + \frac{\partial u}{\partial x_1} v_1 + \frac{\partial u}{\partial x_2} v_2 + \frac{\partial u}{\partial x_3} v_3 \\ &= \frac{\partial u}{\partial t} + \nabla u \cdot \mathbf{v}, \end{aligned}$$

where ∇u is the gradient of u.

2.4.6.1 Conservation of Mass: The Continuity Equation

A balance law in Continuum Mechanics is said to be a *conservation law* if the production p and the physical flux $\mathbf{f}_p$ vanish identically.[9] The content u in this case is to be identified with the *mass density* $\rho = \rho(x_1, x_2, x_3, t)$. Applying the generic equation of balance (2.48) we obtain

$$\frac{\partial \rho}{\partial t} + \operatorname{div}(\rho \mathbf{v}) = 0, \tag{2.50}$$

or, using the material derivative,

$$\frac{D\rho}{Dt} + \rho \operatorname{div}\mathbf{v} = 0. \tag{2.51}$$

This equation is known in Fluid Mechanics as the *continuity equation*.

2.4.6.2 Balance of Linear Momentum

In this case, according to Newton's second law, the content density is the vector of linear momentum, namely, $\rho\, \mathbf{v}$. The production is the *body force* $\mathbf{b}$ per unit spatial volume and the physical flux is the *surface traction vector* $\mathbf{t}$. We will implement the generic balance equation component by component in a global Cartesian inertial frame. For each component t_i of the traction vector, according to Cauchy's theorem, there exists a flux vector with components σ_{ij} $(j = 1, 2, 3)$. We have thus a matrix representing, in the given coordinate system, the components of the *Cauchy stress tensor* $\boldsymbol{\sigma}$. The surface traction $\mathbf{t} = \boldsymbol{\sigma} \cdot \mathbf{n}$ is best expressed in components as

$$t_i = \sigma_{ij} n_j, \tag{2.52}$$

where the *summation convention* for repeated indices has been enforced, as explained in Box 2.2. The equation of balance (2.48) for ρv_i reads

$$\frac{\partial \rho v_i}{\partial t} - b_i - \left(-\rho v_i v_j + \sigma_{ij}\right)_{,j} = 0. \tag{2.53}$$

On the other hand, the continuity equation (2.50) is written in Cartesian components (always enforcing the summation convention) as

$$\frac{\partial \rho}{\partial t} + \left(\rho v_j\right)_{,j} = 0. \tag{2.54}$$

[9] This is the case of the conservation of mass in conventional Continuum Mechanics. In the context of *growing bodies* (such as is the case in some biological materials) mass is not necessarily conserved.

Combining the last two results, we obtain

$$\rho\frac{\partial v_i}{\partial t} + \rho v_{i,j} v_j = b_i + \sigma_{ij,j}. \tag{2.55}$$

Using the material derivative, we can write the local form of the balance of linear momentum as

$$\rho\frac{Dv_i}{Dt} = b_i + \sigma_{ij,j}. \tag{2.56}$$

The material derivative of the velocity is, not surprisingly, the acceleration. Having thus enforced the conservation of mass, the form of Newton's second law for a continuum states that the mass density times the acceleration equals the body force plus the net contact force over the boundary of an elementary volume element.

Box 2.2 The summation convention in Cartesian coordinates

Attributed to Albert Einstein, the summation convention is a notational device which, in addition to being compact and elegant, is often revealing of possible mistakes in computations. In an expression or equation made up of sums of monomials, where each monomial consists of products of indexed quantities, the following rules apply:

1. In each monomial an index (subscript) can appear at most twice. If repeated in a monomial, an index is said to be *dummy*. Otherwise, if it appears just once, it is called *free*.
2. The free indices must be *balanced* in all monomials. In other words, every monomial in an expression must have exactly the same free indices.
3. The dummy indices in every monomial indicate a summation over that index in the range 1 to n, where n is the dimension of the Cartesian space under consideration.

As an example, the divergence of a vector field $\mathbf{v}$ is given by

$$\text{div}\mathbf{v} = v_{i,i},$$

where, as in Eq. (2.2), commas stand for partial derivatives with respect to the coordinates designated by the indices following the comma. An equation such as

$$B_i = A_{ijkjk}$$

stands for

$$B_i = \sum_{k=1}^{n}\sum_{j=1}^{n} A_{ijkjk}.$$

Expressions such as

$$C = D_{kkk} \qquad C_i = D_{jkk} \qquad C_{ij} = D_{ikk}$$

are considered wrong, unless the summation convention has been explicitly suspended.

2.4.6.3 Balance of Angular Momentum

For a continuous deformable medium, unlike the case of a rigid body, the balance of angular momentum is an independent postulate. It establishes that the angular momentum with respect to any point attached to an inertial reference frame is equal to the sum of the moments of the external forces about that point. The angular momentum density is the (pseudo-)vector $\mathbf{r} \times (\rho \mathbf{v})$, where, without any loss of generality, we have identified the fixed point as the origin of coordinates, so that $\mathbf{r}$ is the standard position vector. Assuming the absence of applied couples, and enforcing both mass conservation and balance of linear momentum, the final result is purely algebraic, namely,

$$\sigma_{ij} = \sigma_{ji}. \tag{2.57}$$

Put in words, the Cauchy stress tensor is *symmetric*.

2.4.6.4 Balance of Energy: The First Law of Thermodynamics in a Continuous Medium

The *total energy density* e in a continuum is given by

$$e = \frac{1}{2}\rho \mathbf{v} \cdot \mathbf{v} + \rho \epsilon. \tag{2.58}$$

In this expression, the first term on the right-hand side represents the *kinetic energy density* while ϵ is the *internal energy* per unit mass, postulated to exist as a function of state by the first law of Thermodynamics. The mechanical source density is stipulated by the same law as the *power of the body force*, that is $\mathbf{b} \cdot \mathbf{v}$, while the thermal (that is, non-mechanical) source is provided by sources of heat distributed with a density r per unit volume. Similarly, the physical mechanical flux is given by the power of the traction, that is, $\mathbf{t} \cdot \mathbf{v}$ while the physical heat flux is defined by means of a *heat flux vector* $\mathbf{q}$ such that the non-mechanical influx per unit area and per unit time is given by $-\mathbf{q} \cdot \mathbf{n}$. The balance of energy equates the rate of change of the total energy with the sum of the mechanical and thermal contributions. The final result can be expressed as

$$\rho \frac{D\epsilon}{Dt} = r - q_{i,i} + \sigma_{ij} v_{i,j}, \tag{2.59}$$

where the previous balance equations have been enforced.

Exercises

Exercise 2.1 Derive Eq. (2.18) by applying the equation of balance to a small (infinitesimal) slice of the bar, that is, for a slice contained between the cross sections at x and at $x + dx$.

Exercise 2.2 Carry out all the steps leading to Eq. (2.57) that establishes the symmetry of the Cauchy stress tensor.

Exercise 2.3 Carry out all the steps leading to Eq. (2.59) that establishes the balance of energy in a continuous medium.

Exercise 2.4 (*The Navier-Stokes equations*) The *rate of deformation tensor* **D** is defined as the symmetric part of the velocity gradient. In components

$$D_{ij} = \frac{1}{2}\left(v_{i,j} + v_{j,i}\right).$$

The *Newtonian* compressible viscous fluid has the constitutive equation

$$\sigma = -p(\rho)\mathbf{I} + \lambda\ \mathrm{tr}\mathbf{D} + 2\mu\mathbf{D}.$$

In this equation, $p = p(\rho)$ is some increasing function of the density and λ and μ are constant *viscosity coefficients*. The symbol **I** stands for the spatial identity tensor and the operator tr is the *trace*, namely, $\mathrm{tr}\mathbf{D} = D_{ii}$ is the sum of the diagonal entries in the matrix representing **D** in a Cartesian frame. Use this constitutive equation in the equation of balance of linear momentum to obtain the *Navier-Stokes equations* of Fluid Mechanics [2]. These are three PDEs for the density and the components of the velocity field. The continuity equation completes the system.

References

1. Chadwick P (1999) Continuum mechanics: Concise theory and problems. Dover, New York
2. Chorin AJ, Marsden JE (1993) A mathematical introduction to fluid mechanics. Springer, Berlin
3. Knobel R (2000) An introduction to the mathematical theory of waves. American Mathematical Society, Providence
4. Wolfram S (2002) A new kind of science. Wolfram Media, Champaign

Part II
The First-Order Equation

Chapter 3
The Single First-Order Quasi-linear PDE

Remarkably, the theory of linear and quasi-linear first-order PDEs can be entirely reduced to finding the integral curves of a vector field associated with the coefficients defining the PDE. This idea is the basis for a solution technique known as the method of characteristics. It can be used for both theoretical and numerical considerations. Quasi-linear equations are particularly interesting in that their solution, even when starting from perfectly smooth initial conditions, may break up. The physical meaning of this phenomenon can be often interpreted in terms of the emergence of shock waves.

3.1 Introduction

It has been pointed out by many authors[1] that there is no general theory that encompasses all partial differential equations or even some large classes of them. The exception is provided by the case of a single first order PDE, for which a general theory does exist that, remarkably, reduces the problem to the solution of certain systems of ordinary differential equations. The most general first-order PDE for a function $u = u(x_1, \dots, x_n)$ is of the form

$$F\left(x_1, \dots, x_n, u, u_{,1}, \dots, u_{,n}\right) = 0, \tag{3.1}$$

where F is a function of $2n + 1$ variables.

A commonly used notation, inspired perhaps by Hamiltonian Mechanics, is obtained by defining the new ('momentum') variables

$$p_i = u_{,i} \qquad i = 1, \dots, n. \tag{3.2}$$

The differential equation (3.1) is written as

$$F\left(x_1, \dots, x_n, u, p_1, \dots, p_n\right) = 0. \tag{3.3}$$

[1]This point is made most forcefully by Arnold in [1].

M. Epstein, *Partial Differential Equations*, Mathematical Engineering,
DOI 10.1007/978-3-319-55212-5_3

In a $(2n+1)$-dimensional space with coordinates $x_1, \ldots, x_, u, p_1, \ldots, p_n$ the differential equation acquires a peculiar geometric meaning, briefly described in Box 3.1, which will not be pursued in this book, but which may provide at least a glimmer of a modern geometric way to understand the theory. This approach is pursued in [1].

Box 3.1 A curious geometric view of differential equations
Modern Differential Geometry goes beyond the analysis of curves and surfaces in $\mathbb{R}^3$ and works, instead, on general spaces or *manifolds* of arbitrary dimension that may not even have a metric structure. Although such flights of fancy are well beyond the scope of this book, let us look briefly at the geometrical meaning of a differential equation and its solutions. For simplicity, let us consider the case $n = 1$, namely, the case of an ODE. A PDE is treated similarly in higher dimensional spaces. Consider, therefore, a space with coordinates x, u, p, these coordinates representing, respectively, the independent variable, the dependent variable and its derivative, as shown in Fig. 3.1. Technically, these are coordinates in the *space of jets of dimension 1* over the manifold whose running coordinate is x. In this jet space, the differential equation $F(x, y, p) = 0$ can be regarded as a surface $\mathcal{F}$! More generally, a differential equation is nothing more than a $2n$-dimensional (hyper-)surface embedded in the $(2n + 1)$-dimensional space of 1-jets. This is the geometric meaning of a differential equation. What is a solution? In the case $n = 1$, a solution is a differentiable curve in the x, u space with the property that the derivative $p = du/dx$ at each point x, u determines a point (x, u, p) lying on the differential equation $F = 0$ (that is, the surface $\mathcal{F}$). This analytic statement can also be given a geometric meaning. For any differentiable curve γ in the two-dimensional space x, u we can define its *canonical lift* as a curve γ_* in the three-dimensional jet space x, u, p by simply adding the 'slope' coordinate $p = du/dx$. Thus, a straight line in the space x, u is lifted canonically to a constant ('horizontal') straight line in the jet space. In this view, a solution of the differential equation $F(x, u, p) = 0$ is a curve γ whose canonical lift lives in the surface $\mathcal{F}$.

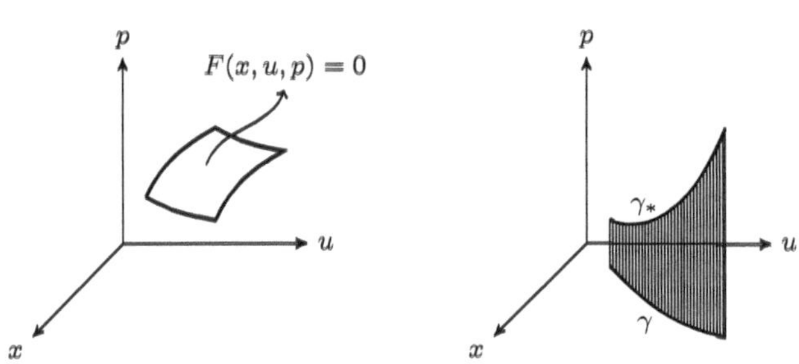

Fig. 3.1 A differential equation (*left*) and the canonical lift of a curve (*right*)

The first-order PDE (3.3) is *quasi-linear* if it is linear in each of the derivative variables p_i $(i = 1, \ldots, n)$. The general form of a quasi-linear first-order PDE is, accordingly,

$$a_1(x_1, \ldots, x_n, u)\frac{\partial u}{\partial x_1} + \cdots + a_n(x_1, \ldots, x_n, u)\frac{\partial u}{\partial x_n} = c(x_1, \ldots, x_n, u). \quad (3.4)$$

If the right-hand side vanishes identically the equation is said to be *homogeneous*. The equation is *linear* if it is linear in both the unknown function u and its derivatives. Thus, a quasi-linear equation is a particular case of a non-linear equation.

3.2 Quasi-linear Equation in Two Independent Variables

In the case of two independent variables, x and y, it is possible to visualize the solution of a PDE for a function $u = u(x, y)$ as a surface[2] in the three-dimensional space with coordinates x, y, u. We call this surface an *integral surface* of the PDE. As pointed out by Courant and Hilbert,[3] geometric intuition is of great help in understanding the theory, so that it seems useful to limit ourselves to the case of two independent variables, at least for now. The visualization of the elements of the theory becomes particularly useful in the case of linear and quasi-linear equations. The general non-linear case is more difficult to grasp and will be the subject of Chap. 5.

The general form of a quasi-linear first-order PDE in two independent variables is

$$a(x, y, u)\, u_x + b(x, y, u)\, u_y = c(x, y, u). \quad (3.5)$$

Consider a possible solution $u = u(x, y)$ of Eq. (3.5). To be considered as a solution it must be at least once differentiable (otherwise, we wouldn't even be able to check that it is a solution).[4] What this means from the geometrical point of view

---

[2]This visualization has nothing to do with the more abstract geometric interpretation given in Box 3.1, which we will not pursue.

[3]In [3], p. 22. This classical treatise on PDEs, although not easy to read, is recommended as a basic reference work in the field of PDEs. A few of the many standard works that deal with first-order PDEs (not all books do) are: [4–7]. Don't be fooled by the age of these books!

[4]Later, however, we will allow certain types of discontinuities of the solution.

is that the surface representing the solution has a well-defined tangent plane at each point of its domain and, therefore, a well-defined normal direction.[5] As we know from Eq. (1.13), at any given point x, y a vector (not necessarily of unit length) in this normal direction is given by the components

$$\begin{Bmatrix} u_x \\ u_y \\ -1 \end{Bmatrix} \tag{3.6}$$

Since Eq. (3.5) can be written as

$$\langle a \ b \ c \rangle \begin{Bmatrix} u_x \\ u_y \\ -1 \end{Bmatrix} = 0, \tag{3.7}$$

we conclude that the statement of the differential equation can be translated into the following geometric statement: The normal to a solution surface must be at each point perpendicular to the *characteristic vector* $\mathbf{w}$ with components $\langle a \ b \ c \rangle$ evaluated at that point in space. But this is the same as saying that this last vector *must lie in the tangent plane to the solution surface!*

Remark 3.1 The use of the imagery of a solution of a first-order PDE in two independent variables as a surface in $\mathbb{R}^3$ is hardly necessary. As discussed in Box 3.2, it carries spurious geometric ideas, such as the normal to the surface. These ideas are certainly useful to visualize the properties of a solution, but may not be directly extended to higher dimensions. That the characteristic vector is tangent to a solution surface can, in fact, be easily proved by purely analytical means. Indeed, let $u = u(x, y)$ be a solution and let $P = (x_0, y_0, u_0 = u(x_0, y_0))$ be a point of this solution. The characteristic vector at this point has components $a_0 = a(x_0, y_0, u_0), b_0 = b(x_0, y_0, u_0), c_0 = c(x_0, y_0, u_0)$. For a nearby point $P + dP = (x_0 + dx, y_0 + dy, u_0 + du)$ to lie in the solution, the increments must satisfy the algebraic relation

$$du = u_x dx + u_y dy,$$

where the derivatives are evaluated at P. If we take $dx = a_0$ and $dy = b_0$ we obtain $du = u_x a_0 + u_y b_0 = c_0$, according to the statement of the PDE. We conclude, therefore, that the characteristic vector lies in the tangent space of the solution at P.

[5]But see Box 3.2.

Box 3.2 What does normal mean?
It is clear that a differentiable function $u = u(x, y)$ has a well-defined tangent plane at each point (x_0, y_0). The equation of this plane is given by

$$u - u_0 = p\,(x - x_0) + q(y - y_0),$$

where $u_0 = u(x_0, y_0)$ and where the derivatives $p = u_x, q = u_y$ are evaluated at (x_0, y_0). But what do we mean by a vector perpendicular to this plane? Our coordinates x, y, u may have completely different meanings and units. Thus, x may represent space, y may represent time and u may be a velocity! In this context, what does it mean to measure lengths of vectors or angles between them? In a rigorous theory, where the space of coordinates x, y, u is not endowed with any intrinsic *metric structure* we cannot speak of lengths and angles. Instead, we must distinguish between vectors and co-vectors and the action of the latter over the former. In particular, the gradient operator produces a co-vector, for which no metric structure is necessary. To avoid these refinements, once we have chosen units for each of the coordinates, we assign to each point a unique numerical triple (x, y, u) and measure lengths and angles in the ordinary sense of $\mathbb{R}^3$. We must, however, make sure that our conclusions are independent of this choice. As a simple example, suppose that we find two vectors each of which is perpendicular to the tangent plane to the surface at some point. We conclude that these two vectors must be collinear. This conclusion is independent of the metric that we have chosen on the basis of the coordinate units.

Let us consider the following problem in the theory of systems of ODEs: Find integral curves of the characteristic vector field $\mathbf{w}(x, y, u)$. From Chap. 1, we have some experience in this type of problem, so we translate it into the system of *characteristic equations*

$$\frac{dx}{ds} = a(x, y, u), \tag{3.8}$$

$$\frac{dy}{ds} = b(x, y, u), \tag{3.9}$$

$$\frac{du}{ds} = c(x, y, u). \tag{3.10}$$

As we know from the theory of ODEs, this system always has a solution (at least locally). This solution can be visualized as a family of non-intersecting integral curves in space. In the context of the theory of first-order quasi-linear PDEs these curves are called the *characteristic curves of the differential equation*, or simply *characteristics*. We have already called the vector field $\mathbf{w}$ with components $\langle a\ b\ c\rangle$ the *characteristic vector field*. The characteristics are, thus, the integral curves of the characteristic vector field.

3.3 Building Solutions from Characteristics

3.3.1 A Fundamental Lemma

Lemma 3.1 *A characteristic curve having one point in common with an integral surface of a quasi-linear first-order PDE necessarily lies on this surface entirely.*

Proof Let $P = (x_0, y_0, u_0)$ be a point lying on an integral surface $u = u(x, y)$ of the PDE and let γ be the unique characteristic curve through P. As a solution of the system of ODEs (3.8), (3.9) and (3.10), γ is expressed by some functions $\hat{x}(s), \hat{y}(s), \hat{u}(s)$. Let s_0 be the value of the parameter s at P. Consider the following function of s

$$U(s) = \hat{u}(s) - u\left(\hat{x}(s), \hat{y}(s)\right). \tag{3.11}$$

We claim that this function vanishes identically over the domain of existence of the characteristic. The geometrical meaning of the function $U(s)$ is displayed in Fig. 3.2. We first project γ onto the (x, y) plane and obtain the *projected characteristic* β as the curve with equations $\hat{x}(s), \hat{y}(s), 0$. Next, we *lift* β to the integral surface as the

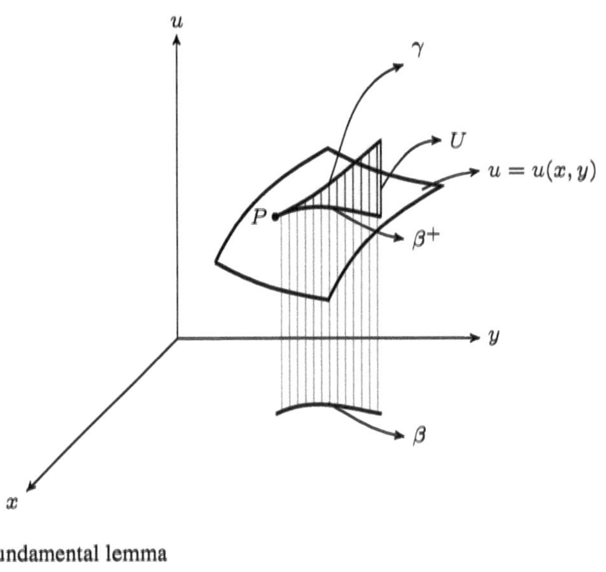

Fig. 3.2 The fundamental lemma

curve β^+ with equations $\hat{x}(s), \hat{y}(s), u\left(\hat{x}(s), \hat{y}(s)\right)$. Our claim is that $\gamma = \beta^+$. The function U is precisely the difference in the third coordinate of these two curves.

Clearly, by definition,

$$U(s_0) = 0. \tag{3.12}$$

A careful calculation of the derivative of U (using the chain rule and enforcing the original differential equation) reveals that

$$\frac{dU}{ds} = \frac{d\hat{u}}{ds} - \frac{\partial u}{\partial x}\frac{d\hat{x}}{ds} - \frac{\partial u}{\partial y}\frac{d\hat{y}}{ds} = \frac{d\hat{u}}{ds} - a\,\frac{\partial u}{\partial x} - b\,\frac{\partial u}{\partial y} = \frac{d\hat{u}}{ds} - c = 0. \tag{3.13}$$

The (unique) solution to this trivial equation with the initial condition (3.12) yields

$$U(s) \equiv 0. \tag{3.14}$$

This result means that γ must lie entirely on the solution surface. □

Remark 3.2 We note that in the proof of this lemma we have not invoked the fact that the characteristic vector at P is tangential to the integral (solution) surface. Put differently, we could have proved the lemma directly from the notion of integral curves of the characteristic field. The fact that the characteristic vectors are tangential to integral surfaces can be obtained as a corollary to the lemma.

3.3.2 *Corollaries of the Fundamental Lemma*

Two important corollaries can be obtained almost immediately from the Fundamental Lemma 3.1.

1. If two different solution surfaces have one point in common, they must intersect along the whole characteristic passing through that point. Conversely, if two solutions intersect *transversely* along a curve, this curve must be a characteristic of the differential equation. By transversely we mean that along the common curve they don't have the same tangent plane. This means that at any point along this curve we have a well-defined line of intersection of the local tangent planes of the two surfaces. But recall that the tangent to the characteristic curve must belong to both tangent planes, and therefore to their intersection. In other words, the intersection between the two planes is everywhere tangent to the characteristic direction and, therefore, the intersection curve is an integral curve of the characteristic field.
2. An integral surface is necessarily a collection of integral curves, since once it contains a point it must contain the whole characteristic through that point. Conversely, any surface formed as a one-parameter collection of characteristic curves is an integral surface of the PDE. What we mean by "one-parameter collection" is that, since the characteristic curves are already one-dimensional entities, to form a surface (which is two-dimensional) we have one degree of freedom left. For

example, we can take an arbitrary non-characteristic line and consider the surface formed by all the characteristics emerging from this line. To show that a surface so formed must necessarily be an integral surface, that is, a solution of the PDE, we consider an arbitrary point P on the given surface. Since, by construction, this surface contains the characteristic through P, the normal to the surface is also perpendicular to the characteristic direction. But this is precisely the statement of the PDE, which is, therefore, satisfied at each point of the given surface.

The general conclusion is that the solutions of a single first-order quasi-linear PDE in two variables can be boiled down to the solution of a system of ordinary differential equations. This result remains true for more than two independent variables and also for fully nonlinear equations (in which case the concept of characteristic curves must be extended to the so-called characteristic strips).

3.3.3 The Cauchy Problem

The main problem in the theory of first-order PDEs is the following so-called *Cauchy problem* or *initial value problem*[6]: Given the values of u on a curve in the x, y plane, find a solution that attains the prescribed values on the given curve. An equivalent way to look at this problem is to regard the given ("initial") data as a space curve with parametric equations

$$x = \bar{x}(r), \tag{3.15}$$

$$y = \bar{y}(r), \tag{3.16}$$

$$u = \bar{u}(r). \tag{3.17}$$

The Cauchy problem consists of finding an integral surface that contains this curve. From the results of the previous section we know that this integral surface must consist of a one-parameter family of characteristics. Let the characteristics be obtained (by integration of the characteristic equations) as

$$x = f(s, A, B, C), \tag{3.18}$$

$$y = g(s, A, B, C), \tag{3.19}$$

$$u = h(s, A, B, C), \tag{3.20}$$

where s is a parameter and where A, B, C are constants of integration. We will adjust these "constants" so that for some fixed value of the parameter ($s = 0$, say) and for

[6]Some authors reserve the name of initial value problem for the particular case in which the data are specified on one of the coordinate axes (usually at t=0).

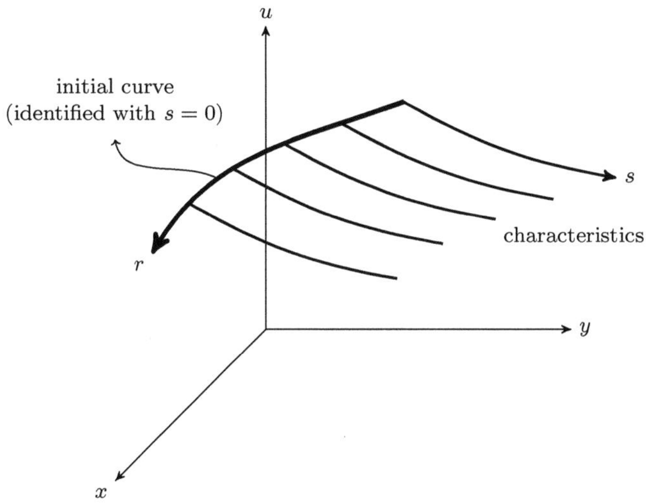

Fig. 3.3 A one-parameter (r) family of characteristics

a given value of r the values of x, y, u given by both sets of equations coincide. In this way we obtain the characteristic issuing from the point r of the initial curve and, moreover, we have adjusted matters so that at the point of their intersection the value of the parameter on the characteristic curve is 0, as suggested in Fig. 3.3. We are in fact creating a coordinate system r, s on the surface being constructed.

This adjustment process leads to the algebraic equations

$$f(0, A, B, C) = \bar{x}(r), \tag{3.21}$$

$$g(0, A, B, C) = \bar{y}(r), \tag{3.22}$$

$$h(0, A, B, C) = \bar{u}(r). \tag{3.23}$$

These equations are, in principle, solvable for A, B, C in terms of r. In other words, the constants of integration are adjusted at each point of the initial curve. This is what we meant by a "one-parameter family of characteristics". We thus obtain (perhaps implicitly) three functions $A(r)$, $B(r)$, $C(r)$. Introducing these functions into the characteristic equations, we obtain finally

$$x = f\,(s, A(r), B(r), C(r)) = \tilde{x}(r, s), \tag{3.24}$$

$$y = g\,(s, A(r), B(r), C(r)) = \tilde{y}(r, s), \tag{3.25}$$

$$u = h\,(s, A(r), B(r), C(r)) = \tilde{u}(r, s), \tag{3.26}$$

These three equations constitute the parametric representation of a surface. Nevertheless, we still have to be able to express it in the form $u = u(x, y)$. In other words, we have to be able to read off the pair r, s in terms of the pair x, y from the first two equations and introduce the result into the third equation. This elimination is possible in principle if, and only if, the Jacobian determinant

$$J = \begin{vmatrix} \tilde{x}_r & \tilde{x}_s \\ \tilde{y}_r & \tilde{y}_s \end{vmatrix} \tag{3.27}$$

does not vanish. Note that, by virtue of Eqs. (3.8) and (3.9), we know that on the solution surface $\tilde{x}_s = a$ and $\tilde{y}_s = b$, we can write the determinant as

$$J = b\,\tilde{x}_r - a\,\tilde{y}_r. \tag{3.28}$$

The problem has thus been completely solved, provided $J \neq 0$. The vanishing of this determinant will be later interpreted as the occurrence of a mathematical *catastrophe*. Physically, the solution ceases to be uniquely defined and a *shock wave* is generated. This situation does not develop if the PDE is linear.

3.3.4 What Else Can Go Wrong?

Suppose that the initial data curve (as a curve in the x, y, u space) happens to be characteristic. In that case, when trying to build a one-parameter family of characteristics, we find that we keep getting the same curve (namely, the initial data curve) over and over again, so that a solution surface is not generated. This should not be surprising. We already know, from Sect. 3.3.2, that characteristics are lines of intersection between solutions. Moreover, using the initial data curve (which is now a characteristic) we can build an infinity of distinct one-parameter families of characteristics to which it belongs. In other words, there are infinite solutions that satisfy the initial data. A different way to express this situation (called the *characteristic initial value problem*) is by saying that the PDE in this case does not provide extra information to allow us to come out uniquely from the initial curve. To drive this point further, we note that by providing differentiable data by means of Eqs. (3.15), (3.16) and (3.17), we are also automatically prescribing the derivative of the desired solution in the direction of the curve, namely, $d\bar{u}/dr = c$. On the other hand, by the chain rule at any point of the initial curve and enforcing the PDE, we know that

$$\frac{d\bar{u}}{dr} = u_x \frac{d\bar{x}}{dr} + u_y \frac{d\bar{y}}{dr} = u_x\, a + u_y\, b = c. \tag{3.29}$$

We conclude that the PDE cannot provide us with information in directions other than characteristic ones. The initial data must remedy this situation by giving us information about the derivative of u in another direction.

If the initial data curve is not characteristic over its whole length but happens to be tangential to a characteristic curve at one point, we have a situation that requires special treatment. An extreme case is obtained when the initial curve is not characteristic anywhere but is everywhere tangent to a characteristic curve. In this case, we have an initial curve that is an envelope of characteristics. Again, this case requires special treatment.

A more subtle situation arises when the initial data are self-contradictory. To see the geometrical meaning of this situation, let $P = (x_0, y_0, z_0)$ be a point on the initial data curve ρ. From the theorem of existence and uniqueness of ODEs, we know that there is a unique characteristic γ through P in some neighbourhood of the point. Assume, moreover, that ρ and γ are not mutually tangent at P. We don't seem to have a problem. But suppose now that ρ and γ project on exactly *the same curve* in the x, y plane. Since the tangent plane of the integral surface at P must contain both the tangent to γ (because it is the local characteristic vector) and the tangent to ρ (because the initial curve must belong to the solution), we obtain a vertical tangent plane, which is not permitted if u is differentiable.

3.4 Particular Cases and Examples

3.4.1 Homogeneous Linear Equation

Consider the homogeneous linear equation

$$a(x, y)\, u_x + b(x, y)\, u_y = 0. \tag{3.30}$$

By homogeneous we mean that the right-hand side (the term without derivatives) is zero. It follows immediately from the characteristic equations (3.8), (3.9) and (3.10) that the first two equations can be integrated separately from the third. What this means is that we can now talk about characteristic curves in the x, y plane. From Eq. (3.10), we see that the value of u on these "projected" characteristic curves must be constant. In other words, the original characteristic curves are contained in horizontal planes and they project nicely onto non-intersecting curves in the x, y plane.

Example 3.1 (*Advection equation*) Find the solution of the following advection equation with constant coefficients

$$u_t + 3u_x = 0, \tag{3.31}$$

with the initial condition

$$u(x, 0) = \frac{1}{x^2 + 1}. \tag{3.32}$$

Solution: The characteristic curves are given by the solutions of the system

$$\frac{dt}{ds} = 1 \qquad \frac{dx}{ds} = 3 \qquad \frac{du}{ds} = 0. \tag{3.33}$$

This system is easily integrated to yield

$$t = s + A \qquad x = 3s + B \qquad u = C. \tag{3.34}$$

The initial curve in this case lies on top of the x axis, so we can choose x itself as a parameter. To preserve the notation of the general procedure, we write the equation of the initial curve explicitly as

$$t = 0 \qquad x = r \qquad u = \frac{1}{r^2 + 1}. \tag{3.35}$$

Now is the time to enforce Eqs. (3.21), (3.22) and (3.23) to obtain

$$A = 0 \qquad B = r \qquad C = \frac{1}{r^2 + 1}. \tag{3.36}$$

Introducing this result into Eq. (3.34), we get

$$t = s \qquad x = 3s + r \qquad u = \frac{1}{r^2 + 1}. \tag{3.37}$$

We still need to eliminate the pair r, s in favour of x, y, an operation which in this case is trivial. We introduce the result into the last expression in (3.37) and obtain

$$u = \frac{1}{(x - 3t)^2 + 1}. \tag{3.38}$$

This is the desired solution. It consists of a traveling wave of the same shape as the initially prescribed profile. This is precisely the physical meaning of the advection equation with constant coefficients. The wave travels forward with a speed of 3.

Remark 3.3 When producing an exact solution of a PDE, it is a good idea at the end of the whole process to verify by direct substitution that the proposed solution satisfies the given equation and the initial conditions.

3.4.2 Non-homogeneous Linear Equation

The general form of the non-homogeneous linear equation is

$$a(x, y)\, u_x + b(x, y)\, u_y = c(x, y)\, u + d(x, y). \tag{3.39}$$

The characteristic equations are

$$\frac{dx}{ds} = a(x, y), \tag{3.40}$$

$$\frac{dy}{ds} = b(x, y), \tag{3.41}$$

$$\frac{du}{ds} = c(x, y)\, u + d(x, y). \tag{3.42}$$

Just as in the case of the homogeneous equation, we observe that the first two equations can be solved independently to yield a family of non-intersecting curves in the x, y plane. The value of u on these lines, however, is no longer constant. Again, the characteristic curves project nicely on the x, y plane, since the third equation can be solved on the basis of the first two, curve by curve. Notice that from this point of view, the linearity of the right-hand side doesn't play a determining role. The method of solution for the non-homogeneous equation follows in all respects the same lines as the homogeneous one.

Example 3.2 (*Propagation of discontinuities*) A first-order PDE admits discontinuous initial data. The reason for this somewhat surprising fact is precisely that the construction of the solution shows that the initial data are propagated along characteristics. In the linear case, the projected characteristics do not interact with their neighbours. Recall also that, as remarked in Sect. 3.3.4, the PDE prescribes derivatives in the characteristic directions only, so that the fact that there is a discontinuity in the transverse direction does not impair the verification of the validity of the discontinuous solution. As an example, consider the linear PDE

$$u_t + 3u_x = -2u\, x, \tag{3.43}$$

with the initial condition

$$u(x, 0) = H[x], \tag{3.44}$$

where $H[x]$ is the Heaviside (unit step) function.

Solution:

The initial curve is given in parametric form by

$$t = 0 \qquad x = r \qquad u = H[r]. \tag{3.45}$$

The characteristic differential equations are

$$\frac{dt}{ds} = 1 \qquad \frac{dx}{ds} = 3 \qquad \frac{du}{ds} = -2u\, x. \tag{3.46}$$

This system is easily integrated to

$$t = s + A \qquad x = 3s + B \qquad u = C\, e^{-3s^2-2Bs}. \tag{3.47}$$

Setting $s = 0$ and equating to the initial curve equations, we obtain:

$$A = 0 \qquad B = r \qquad C = H[r]. \tag{3.48}$$

Putting all these results together, just as in the previous example, we obtain the solution as

$$u = H[x - 3t]\, e^{3t^2-2xt}. \tag{3.49}$$

We observe that the discontinuity propagates along the corresponding characteristic. Nevertheless, since the characteristics in this case are not horizontal as before, the magnitude of the jump changes progressively. In this case, on one side of the projected characteristic (where $x < 3t$) the solution vanishes identically, whereas on the other side it has the value e^{-3t^2}, which is thus the value of the jump. In this case, the discontinuity gets attenuated as time progresses.

3.4.3 *Quasi-linear Equation*

So far, everything seems to be working smoothly. The problems start once one crosses the threshold into the non-linear realm. For the time being, nonlinear for us means quasi-linear, since we have not dealt with the genuinely nonlinear case. Geometrically speaking, the reason for the abnormal behaviour that we are going to observe is that in the case of a quasi-linear equation the characteristics (that live in the 3-dimensional space x, y, u) do not project nicely onto the plane of the independent variables x, y. From the point of view of the characteristic system of ODEs, this is the result of the fact that the first two equations are coupled with the third, unlike the case of the linear equation, where u was not present in the first two equations. As a consequence of this coupling, for the same values of x, y, but for a different value of u, we obtain, in general, characteristics that do not project onto the same curve. In other words, the projected characteristics intersect and the projected picture is a veritable mess. Figures 3.4 and 3.5 may help to understand the above statements. Therefore, given smooth initial conditions on a curve may lead to intersections of the projected characteristics emerging from the initial curve. What this means is that at one and the same point in the space of independent variables, we may end up having two different solutions. When the independent variables are space and time, this situation is usually described as the development of a shock after a finite lapse of time. This mathematical catastrophe is accompanied by physical counterparts (sonic booms, for example).

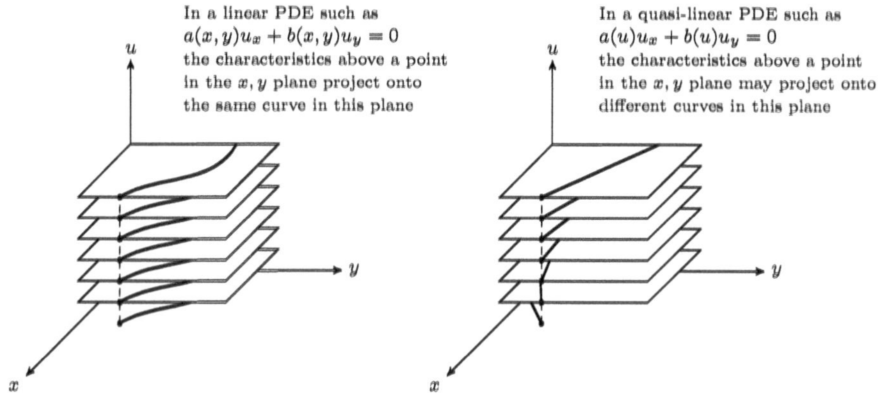

Fig. 3.4 Characteristics of linear and quasi-linear equations

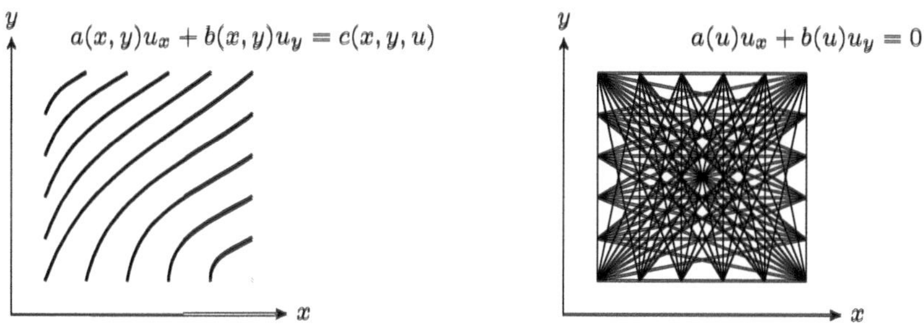

Fig. 3.5 Projected characteristics

Example 3.3 (*Breaking of waves*) Consider the quasi-linear PDE

$$u_t + u\, u_x = 0. \tag{3.50}$$

This equation is known as the *inviscid Burgers equation*. It has important applications to gas dynamics.

The characteristic ODEs of Eq. (3.50) are

$$\frac{dt}{ds} = 1 \qquad \frac{dx}{ds} = u \qquad \frac{du}{ds} = 0, \tag{3.51}$$

which can be integrated to obtain the characteristic curves as

$$t = s + A \qquad x = C\, s + B \qquad u = C. \tag{3.52}$$

Let us regard pictorially the solutions $u = u(x, t)$ as waves which at any given time τ have the geometric profile given by the function of the single variable x defined as $u = u_{(\tau)}(x) = u(x, \tau)$. This way of looking at the solution usually corresponds to the physical interpretation and makes it easier to describe what is going on in words. We will now consider initial data at time $t = 0$, that is, we will prescribe a certain initial profile given by some function $f(x)$, namely,

$$u(x, 0) = u_{(0)}(x) = f(x), \tag{3.53}$$

where $f(x)$ is assumed to be smooth. We will investigate what happens thereafter (for $t > 0$).

Consider first the easy case in which $f(x)$ is a monotonically increasing function of x, such as

$$f(x) = \tanh(x). \tag{3.54}$$

The initial curve for our Cauchy problem is then given parametrically by

$$t = 0 \qquad x = r \qquad u = \tanh(r). \tag{3.55}$$

Following the general procedure (equating the characteristic expressions for $s = 0$ with the corresponding equations for the initial curve), we obtain

$$A = 0 \qquad B = r \qquad C = \tanh(r). \tag{3.56}$$

We have thus obtained the following one-parameter family of characteristics, where the parameter of the family is r:

$$t = s \qquad x = s\ \tanh(r) + r \qquad u = \tanh(r). \tag{3.57}$$

Recall that r represents the running coordinate in the initial curve (along the x axis). The projection onto the plane x, t of these characteristics is given parametrically by the first two expressions above. In this case, it is easy to eliminate the characteristic parameter s and to write the family of projected characteristics as

$$x = t\ \tanh(r) + r. \tag{3.58}$$

For each r, this is the equation of a straight line. The situation is represented in Fig. 3.6 (produced by Mathematica®), which shows that the projected characteristics fan out, as it were, and that they will never intersect for $t > 0$.

We now ask: What happens to the wave profile as time goes on? Since our solution (if it exists) is given parametrically by Eq. (3.57), the profile for a fixed time $\tau > 0$ is given parametrically by

$$x = \tau\ \tanh(r) + r \qquad u = \tanh(r). \tag{3.59}$$

```
Show[Table[ParametricPlot[{s*Tanh[0.1*i]+0.1*i,s},{s,0,2}, PlotRange->{{-1,1},{0,2}},
    FrameLabel->{x,t}, Frame->True, PlotStyle->Black], {i,-9,9}]]
```

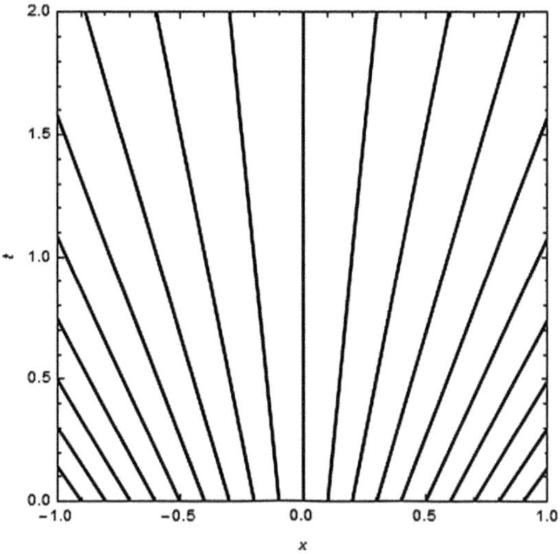

Fig. 3.6 Characteristics fanning out towards the future

Let us plot this profile separately for three different times $\tau = 0, 2, 4$, as shown in Fig. 3.7.

The wave profile tends to smear itself out over larger and larger spatial extents. We can now plot the solution as a surface in a three-dimensional space with coordinates x, t, u (for $t \geq 0$), as shown in Fig. 3.8.

Consider now instead an initial profile with rising and falling parts, such as

$$f(x) = \frac{1}{x^2 + 1}. \tag{3.60}$$

```
GraphicsRow[Table[ParametricPlot[{i*Tanh[r]+r,Tanh[r]},{r,-7.5,7.5}, PlotRange->{{-7,7},{-1.2,1.2}},
    AxesLabel->{x,u}, PlotStyle->Black, AspectRatio->0.8],{i,1,3,1}]]
```

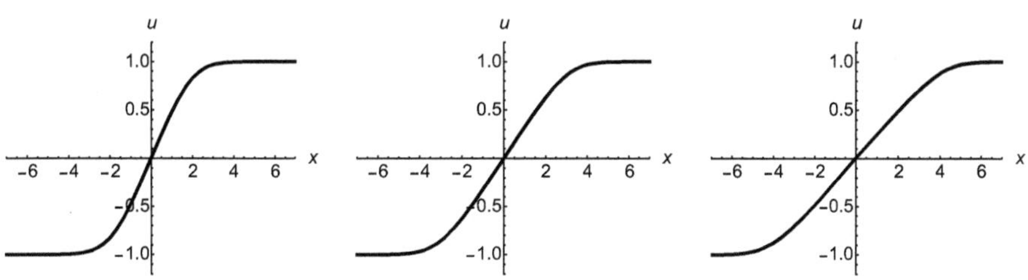

Fig. 3.7 Wave profiles at times $t = 1, 2, 3$

```
ParametricPlot3D[{s,s*Tanh[r]+r,Tanh[r]},{s,0,10},{r,-15,15}, AxesLabel->{t,x,u},
Mesh->25, BoxRatios->{1,2,1}, PlotRange->{{0,10},{-15,15}}]
```

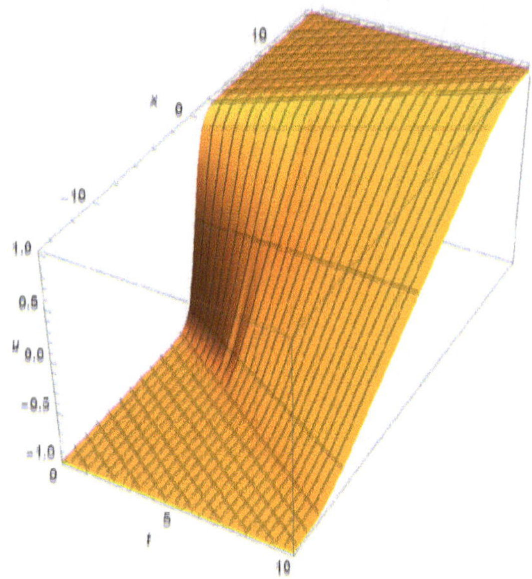

Fig. 3.8 Graph of the solution

The characteristics are obtained as

$$t = s \qquad x = \frac{s}{r^2+1} + r \qquad u = \frac{1}{r^2+1}. \tag{3.61}$$

For each value of r, the projected characteristic is the straight line

$$x = \frac{t}{r^2+1} + r, \tag{3.62}$$

but now the slope is always positive, with a pattern that fans out or in for positive or negative r, respectively. Figure 3.9 shows the state of affairs.

From Eq. (3.61) we can read off for any $t = \tau$ the following parametric relation between u and x:

$$x = \frac{\tau}{r^2+1} + r \qquad u = \frac{1}{r^2+1}. \tag{3.63}$$

Figure 3.10 shows the wave profiles for $\tau = 0.0, 0.5, 1.0, 1.5, 2.0, 2.5$.

```
Show[Table[ParametricPlot[{s/(l^2+1)+l,s},{s,0,4},PlotRange->{{-3,4},{0,4}}, FrameLabel->{x,t},
    Frame->True, PlotStyle->Black],{l,-3,4,0.2}]]
```

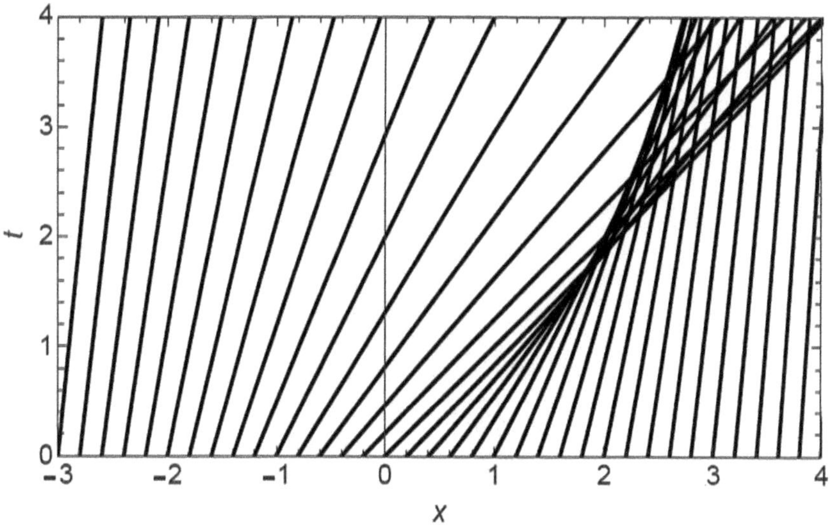

Fig. 3.9 Characteristics intersect forming an envelope with a cusp

We see that as two projected characteristics converge, the corresponding points of the profile become closer and closer in projection (that is, more and more on top of each other), so that the slope of the profile tends to infinity. Although there is no problem drawing the parametric profiles, we see that, in fact, the solution ceases to exist at that point in time in which the slope of the profile becomes infinite. We have a catastrophe, like the breaking of a water wave. The reason why we say that the solution ceases to exist is that we can no longer express u as a (single-valued) function of x and y. Indeed, in order to be able to do so, we should be able, from the first two Eq. (3.61), namely,

$$t = s \qquad x = \frac{s}{r^2+1} + r, \tag{3.64}$$

to read off r and s as functions of x and t. For this to be possible, we need the Jacobian determinant

$$J = \begin{vmatrix} \frac{\partial x}{\partial r} & \frac{\partial x}{\partial s} \\ \frac{\partial t}{\partial r} & \frac{\partial t}{\partial s} \end{vmatrix} = \begin{vmatrix} \frac{-2r\,s}{r^2+1} + 1 & \frac{1}{r^2+1} \\ 0 & 1 \end{vmatrix} \tag{3.65}$$

not to vanish. The vanishing of J occurs when

$$\left(r^2+1\right)^2 - 2r\,s = 0. \tag{3.66}$$

```
Show[Table[ParametricPlot[{r+i/(r^2+1),1/(r^2+1)},{r,-10,10}, PlotRange->{{-7,7},{-0,1.2}},
    AxesLabel->{x,u}, PlotStyle->Black, AspectRatio->0.8, PlotPoints->20],{i,0,3.5,0.7}]]
```

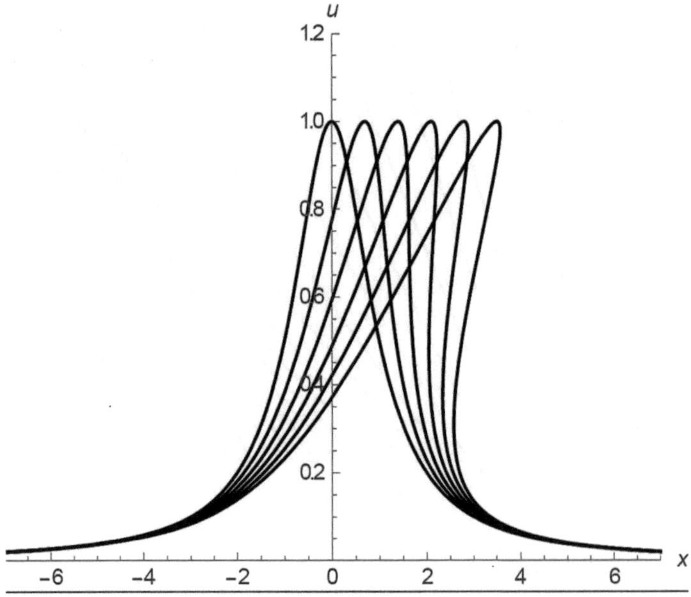

Fig. 3.10 Wave profiles at times $t = 0, 0.7, 1.4, 2.1, 2.8, 3.5$

Of all these combinations of r and s we are looking for that one that renders the minimum value of t. In general, t is a function of both r and s, so that we have to solve for the vanishing differential of this function under the constraint (3.66). In our particular example, since $t = s$, this is a straightforward task. We obtain

$$dt = ds = d\left(\frac{\left(r^2+1\right)^2}{2r}\right) = 0. \tag{3.67}$$

Expanding this expression and choosing the smallest root we obtain the value of r at the breaking point as

$$r_b = \frac{\sqrt{3}}{3}, \tag{3.68}$$

which yields the breaking time

$$t_b = s_b = \frac{\left(r_b^2+1\right)^2}{2r_b} = \frac{8\sqrt{3}}{9} \approx 1.54. \tag{3.69}$$

This value is corroborated by a look at the graph of the intersecting projected characteristics. An equivalent reasoning to obtain the breaking time could have been the following. Equation (3.63) provides us with the profile of the wave at time $t = \tau$.

We are looking for the smallest value of τ that yields an infinite slope du/dx. This slope is calculated as

$$\frac{du}{dx} = \frac{du/dr}{dx/dr} = \frac{2r}{2r\,\tau - \left(r^2+1\right)^2}. \tag{3.70}$$

For it to go to infinity, the denominator must vanish, whence

$$\tau = \frac{\left(r^2+1\right)^2}{2r}. \tag{3.71}$$

The minimum time is obtained by equating $d\tau/dr$ to zero, thus yielding the same result as Eq. (3.68).

Once we have calculated the breaking time t_b we can determine the spatial location x_b of the shock initiation. It is obtained from Eq. (3.64) as

$$x_b = \frac{s_b}{r_b^2+1} + r_b = \sqrt{3}, \tag{3.72}$$

where we have used Eqs. (3.64) and (3.68).

We have learned from this example that, in general, a quasi-linear first-order PDE may develop breaking catastrophes of the type just described. The solution obtained by integrating the characteristic equations and exhibiting a one-parameter family thereof is not valid beyond the breaking time (which can, however, be calculated by that very method). In physical applications, such as gas dynamics, this event is interpreted as the birth of a shock wave. It is important to notice that the shock develops out of *perfectly smooth initial conditions*. Anyone doubting the power of mathematics to predict, at least in principle, real world events should take a close look at this example.[7] Physics, however, does not come to a halt simply because a certain mathematical singularity has been detected. The phenomena continue their inexorable path. In the next chapter we will investigate how the mathematical problem can be resolved.

3.5 A Computer Program

The method of characteristics can be used efficiently for the numerical solution of first-order PDEs. Indeed, all one needs to do is to program the initial value problem for a system of first-order ODEs. In the MATLAB® code shown in Box 3.3 an elementary (though not very accurate) forward difference method is used. The accuracy can be improved by replacing this integrator by the Runge–Kutta method. The user must input the functions describing the coefficients of the PDE. The case illustrated corresponds to the inviscid Burgers equation solved above. Figure 3.11 shows the graph of the numerical solution produced by the code.

[7] The classical reference work for this kind of problem is [2]. The title is suggestive of the importance of the content.

Box 3.3 A simple computer code

```
function CharacteristicNumMethod
% The quasi-linear PDE aa*u_x + bb*u_t= cc is solved by the method
% of characteristics. Users must specify the functions aa, bb and cc.
% The solution is graphed in parametric form, so that shock formation
% can be discerned from the graph. The shock is not treated. Initial
% conditions are specified by function init at time t=0
N = 50; % Number of steps along the parameter s
M = 200; % Number of steps along the parameter r
ds = 0.1; % Parameter s step size
dr = 0.05; % Parameter r step size

solution = zeros(N,M,5);

for j = 1:M % Iteration over initial condition parameter 'r'
    x0 = (j-M/2)*dr;
    t0 = 0;
    u0=init(x0);

for i=1:N % Iteration over curve continuation parameter 's'
    solution(i,j,1) = x0;
    solution(i,j,2) = t0;
    solution(i,j,3) = u0;

    % Forward differences
    x00 = x0+aa(x0,t0,u0)*ds;
    t00 = t0+bb(x0,t0,u0)*ds;
    u00 = u0+cc(x0,t0,u0)*ds;
      x0=x00;
      t0=t00;
      u0=u00;
end
end

% Plot solution
kc=0;
for i=1:N
    for j=1:M
        kc=kc+1;
        XX(i,j)=solution(i,j,1);
        TT(i,j)=solution(i,j,2);
        UU(i,j)=solution(i,j,3);

    end
end

figure(1);

 S = surf(XX, TT, UU);
 xlabel('x'); ylabel('t'); zlabel('u');
    end

function aa = aa(x,t,u)
% Coefficient of u_x
aa = u;
end

function bb = bb(x,t,u)
% Coefficient of u_t
bb = 1;
end

function cc = cc(x,t,u)
% Right-hand side
cc =0;
end

function init=init(x)
init=1/(x^2+1);
end
```

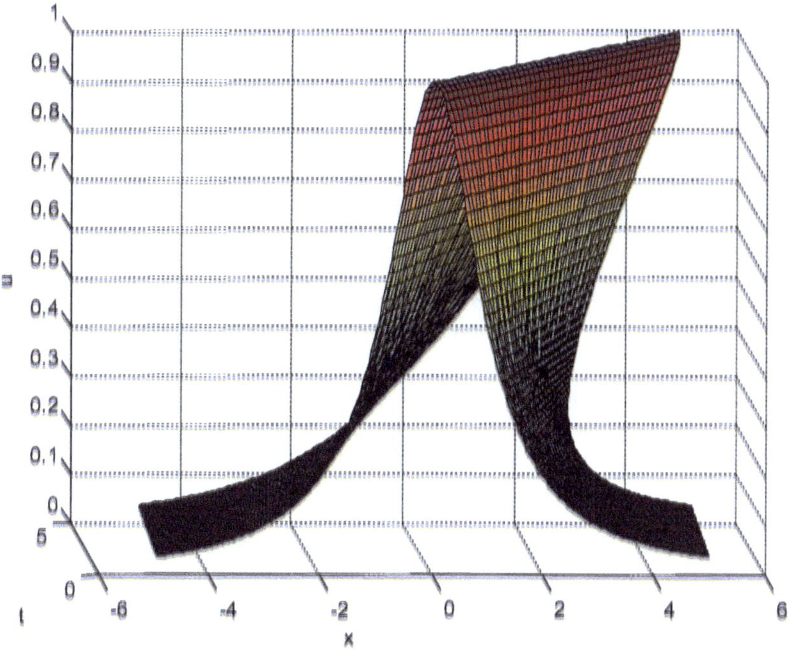

Fig. 3.11 Numerical solution of the inviscid Burgers equation

Exercises

Exercise 3.1 What conditions must the functions a, b, c in (3.5) satisfy for this equation to be linear?

Exercise 3.2 Are the characteristics of a linear equation necessarily straight lines? Are the characteristic ODEs of a linear PDE necessarily linear? Are the characteristic ODEs of a quasi-linear PDE necessarily non-linear?

Exercise 3.3 ([8], *p.* 96) Solve the following initial-value problem for $v = v(x, t)$:

$$v_t + c\, v_x = x\, t \qquad v(x, 0) = \sin x,$$

where c is a constant.

Exercise 3.4 Find the solution of the equation

$$(x - y)u_x + (y - x - u)u_y = u$$

for a function $u = u(x, y)$ passing through the line $y = x,\ u = 1$.

Exercise 3.5 Solve the differential equation

$$y\,\frac{\partial u}{\partial x} + x\,\frac{\partial u}{\partial y} = u \tag{3.73}$$

with the initial condition

$$u(x, 0) = x^3. \tag{3.74}$$

Further questions: Is this a linear PDE? What are the projected characteristics of this PDE? Is the origin an equilibrium position for the characteristic equations? Do the initial conditions determine the solution uniquely for the whole of $\mathbb{R}^2$? What would the domain of existence of the solution be if the initial conditions had been specified on the line $y = 1$ instead of $y = 0$?

Exercise 3.6 Modify the MATLAB® code of Box 3.3 to solve Exercise 3.5 numerically. Apart from the obvious changes to the functions aa, bb, cc and init, you may want to modify some of the numerical values of the parameters at the beginning of the program controlling the range and precision of the solution. Comment on quantitative and qualitative aspects of the comparison between the exact solution obtained in Exercise 3.5 and the numerical solution provided by the program. Does the numerical solution apply to the whole of $\mathbb{R}^2$?

Exercise 3.7 Find the breaking time of the solution of the inviscid Burgers equation (3.50) when the initial condition is given by the sinusoidal wave $u(x, 0) = \sin x$. Where does the breaking occur?

Exercise 3.8 Consider the modified Burgers equation

$$u_t + u^2 u_x = 0.$$

(a) Show that the new function $w = u^2$ satisfies the usual Burgers equation. (b) On the basis of this result find the solution of the initial value problem of the modified Burgers equation with initial condition $u(x, 0) = x$. (c) What is the domain of validity of this solution? (d) Is any simplification achieved by this change of variables?

References

1. Arnold VI (2004) Lectures on partial differential equations. Springer, Berlin
2. Courant R, Friedrichs KO (1948) Supersonic flow and shock waves. Interscience, New York
3. Courant R, Hilbert D (1962) Methods of mathematical physics, vol II. Interscience, Wiley, New York
4. Duff GFD (1956) Partial differential equations. University of Toronto Press, Toronto
5. Garabedian PR (1964) Partial differential equations. Wiley, New York
6. John F (1982) Partial differential equations. Springer, Berlin
7. Sneddon IN (1957) Elements of partial differential equations. McGraw-Hill, Maidenheach Republished by Dover 2006
8. Zauderer E (1998) Partial differential equations of applied mathematics, 2nd edn. Interscience, Wiley, New York

Chapter 4
Shock Waves

When the projected characteristics of a quasi-linear first-order PDE intersect within the domain of interest, the solution ceases to exist as a well-defined function. In the case of PDEs derived from an integral balance equation, however, it is possible to relax the requirement of continuity and obtain a single-valued solution that is smooth on either side of a shock front of discontinuity and that still satisfies the global balance law. The speed of propagation of this front is obtained as the ratio between the jump of the flux and the jump of the solution across the shock.

4.1 The Way Out

The impasse exposed at the end of Chap. 3 can be summarized as follows: For some quasi linear PDEs with given initial conditions the solution provided by the method of characteristics ceases to be single-valued. It may very well be the case that a reinterpretation of the PDE, or a modification of the dependent and independent variables, can allow us to live with the multi-valued nature of the result. On the other hand, in most applications, the function $u = u(x, y)$ is the carrier of an intrinsic physical variable, whose single-valued nature is of the essence.[1] In this event, the engineer may wish to check whether or not the model being used has been obtained by means of excessive simplifications (for example, by neglecting higher-order derivatives). It is remarkable, however, that a way out of the impasse can be found, without discarding the original model, by allowing a generalized form of the equation and its solutions. Indeed, these generalized solutions are only *piecewise* smooth. In simpler words, these solutions are differentiable everywhere, except at a sub-domain of measure zero (such as a line in the plane), where either the function itself or its derivatives

[1]For this and other points in the theory of shocks in one-dimensional conservation laws, references [2, 5] are recommended.

M. Epstein, *Partial Differential Equations*, Mathematical Engineering,
DOI 10.1007/978-3-319-55212-5_4

experience discrete jumps. Moreover, a weaker interpretation of the conservation law itself may be allowed. By *weak* we mean an integral form of the balance laws that admits solutions with a lower degree of differentiability than we do in the differential version of the equations. These generalized and weak solutions, carrying discontinuities born at the breaking time and developing into the future, are what we call *shocks*.

4.2 Generalized Solutions

We recall that, in practice, PDEs more often than not appear in applications as the result of the statement of an equation of balance. In Chap. 2, Eq. (2.14), we learned that the preliminary step towards obtaining the balance PDE for the case of one spatial variable x had the integral form

$$\frac{d}{dt}\int_{x_1}^{x_2} u(x,t)\,dx = \int_{x_1}^{x_2} p(x,t)\,dx + f(x_2,t) - f(x_1,t), \tag{4.1}$$

where x_1, x_2 are arbitrary limits with $x_1 < x_2$. There is no reason, therefore, to discard this integral version in favour of the differential counterpart, since in order to obtain the latter we needed to assume differentiability of $u = u(x,t)$ over the whole domain of interest. This domain will be divided into two parts,[2] namely, the strip $0 \le t \le t_b$ and the half-plane $t > t_b$. The upper part, moreover, will be assumed to be subdivided into two regions, $\mathcal{D}^-$ and $\mathcal{D}^+$, in each of which the solution, u^- and u^+, is perfectly smooth. These two regions are separated by a smooth curve with equation

$$x = x_s(t) \tag{4.2}$$

along which discontinuities in u and/or its derivatives may occur. This curve, illustrated in Fig. 4.1, is, therefore, the *shock front* (the carrier of the discontinuity). The reason that we are willing to accept this non-parametric expression of the shock curve is that, from the physical point of view, the derivative dx_s/dt represents the instantaneous speed of the shock, which we assume to be finite at all times. The shock curve, clearly, passes through the original breaking point, with coordinates (x_b, t_b).

We evaluate the integral on the left-hand side of Eq. (4.1) for times beyond t_b (and for a spatial interval (a, c) containing the shock) as

$$\frac{d}{dt}\int_{a}^{c} u(x,t)\,dx = \frac{d}{dt}\int_{a}^{x_s(t)} u(x,t)\,dx + \frac{d}{dt}\int_{x_s(t)}^{c} u(x,t)\,dx \tag{4.3}$$

---

[2] For the sake of simplicity, we are assuming that no other shocks develop after the breaking time t_b that we have calculated.

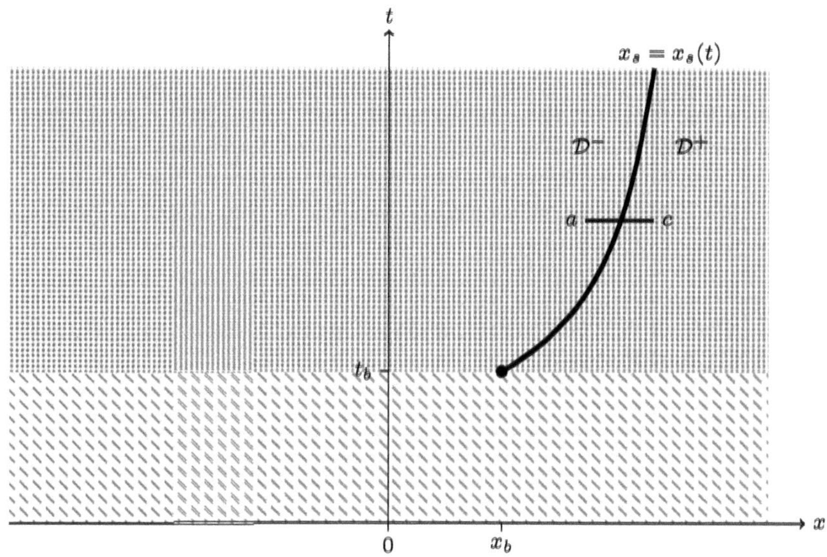

Fig. 4.1 The shock front

Because the limits of these integrals depend on the variable with respect to which we are taking derivatives (t), we can no longer simply exchange the derivative with the integral. Either by doing the derivation yourself, or by consulting a Calculus textbook, you can convince yourself of the following formula[3]

$$\frac{d}{dt}\int_{f(t)}^{g(t)} u(x,t)\,dx = \int_{f(t)}^{g(t)} \frac{\partial u(x,t)}{\partial t}\,dx + u(g(t),t)\frac{dg(t)}{dt} - u(f(t),t)\frac{df(t)}{dt}. \quad (4.4)$$

Introducing this formula into Eq. (4.3) and the result into (4.1), we obtain

$$\int_{a}^{x_s(t)} \frac{\partial u}{\partial t}\,dx + u(x_s^-,t)\frac{dx_s}{dt} + \int_{x_s(t)}^{c} \frac{\partial u}{\partial t}\,dx - u(x_s^+,t)\frac{dx_s}{dt} = f(c,t) - f(a,t), \quad (4.5)$$

where the superscripts "−" and "+" indicate whether we are evaluating the (possibly discontinuous) solution immediately to the left or to the right of the shock curve, respectively. In Eq. (4.5) we have assumed that there is no production, since its presence would not otherwise affect the final result. We now let the limits of the original integral, a and c, approach x^- and x^+, respectively, and obtain the speed of the shock as

[3]The generalization of this formula to three dimensions is known as Reynolds' transport theorem, with which you may be familiar from a course in Continuum Mechanics.

$$\frac{dx_s}{dt} = \frac{f(x_s^+, t) - f(x_s^-, t)}{u(x_s^-, t) - u(x_s^+, t)}. \tag{4.6}$$

This result is known as the *Rankine–Hugoniot jump condition*.[4] It establishes that for a discontinuous (shock) solution to exist, the shock must propagate at a speed equal to the ratio between the jump of the flux and the jump of the solution across the shock curve. This extra condition allows, in general, the solution to be extended beyond the time of shock formation.

4.3 A Detailed Example

Many textbooks present examples of propagation of shocks in which the discontinuity is already imposed in the initial conditions. Moreover, the initial profile is very simple. A typical example, discussed in Sect. 4.4 below, imposes the step function $u(x, 0) = 1 - H[x]$, so that the characteristics consist of two sets of straight parallel lines. The reason behind these idealized examples can be found in the inherent difficulty implied by the integration of the Rankine–Hugoniot condition. Let us consider, however, the case of the PDE (3.50) with the initial condition (3.60), with which we are familiar already, namely

$$u_t + u\, u_x = 0 \qquad u(x, 0) = \frac{1}{x^2 + 1}. \tag{4.7}$$

We have calculated the breaking time t_b for this problem. We are now in a position to do more than this, namely, to determine the whole zone whereby the solution is multi-valued. This information is obtained by implementing the zero-determinant condition (3.66) in Eq. (3.64), which yield the bounding line of this zone in parametric form. The plot of this boundary is shown in Fig. 4.2.

It should not be surprising that this line contains a cusp (at time t_b), since it is obtained as an envelope of characteristic lines which, as shown in Fig. 3.9, turn first one way and then the other. The shock curve starts at this cusp point and develops within the zone enclosed by the two branches according to the Rankine–Hugoniot condition. Notice that at each point within this domain, there are three values produced by the characteristic method. The middle value is irrelevant, since the solution must be smooth on either side of the shock curve, and the Rankine–Hugoniot condition must be enforced between the highest and the lowest value. What this means is that, within the domain of multiplicity, we have a well-defined and smooth right-hand side of the Rankine–Hugoniot condition, which can then be regarded as an ODE for the determination of the shock curve. The only problem is the explicit calculation of the highest and lowest values of the solution.

---

[4]Naturally, because of the sign convention we used for the flux, the formula found in most books changes the sign of the right-hand side.

```
ParametricPlot[{(r^2+1)/(2*r)+r,(r^2+1)^2/(2*r)},{r,0,2}, AxesLabel->{x,t},
    PlotRange->{{-1,5},{0,5}}, PlotStyle->Black]
```

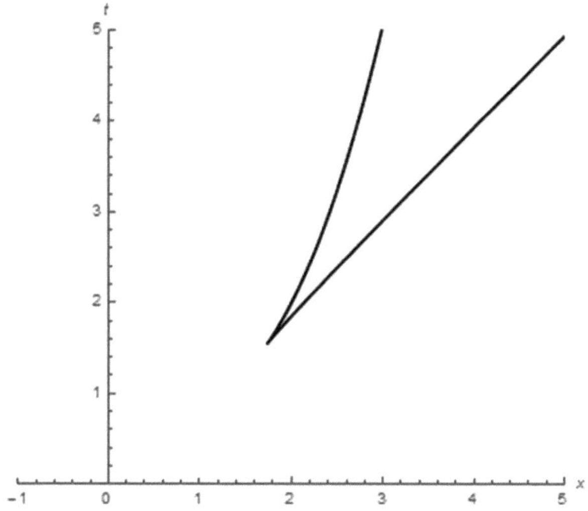

Fig. 4.2 Boundary of the domain of multiplicity

In our particular case, it is not difficult to see, by looking at Eqs. (2.24) and (3.50), that the flux function is given by

$$f(u) = -\frac{1}{2}u^2. \tag{4.8}$$

Introducing this result into Eq. (4.6), we obtain

$$\frac{dx_s}{dt} = \frac{(u^+)^2 - (u^-)^2}{2(u^+ - u^-)} = \frac{1}{2}\left(u^+ + u^-\right). \tag{4.9}$$

This means that the shock curve negotiates its trajectory in such a way as to have a slope equal to the average of the solutions coming from either side. This can be enforced either analytically (if this average is easily available) or numerically (if it isn't). Figure 4.3 illustrates what average we are talking about, by showing a typical profile of the multi-valued solution for some $t > t_b$.

So, for each instant of time $t > t_b$, moving the vertical line between its two extreme points of tangency (which project onto the boundary of the multi-valued zone), we obtain a smooth function for the right-hand side of Eq. (4.9). In our particular example, to obtain the values u^+ and u^- analytically we need to solve for the highest and lowest roots of a cubic equation. Indeed, according to Eq. (3.63), the profile at time τ is given parametrically by

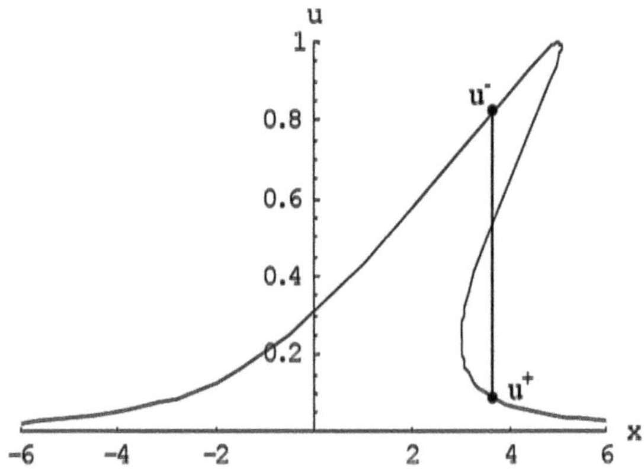

Fig. 4.3 The shock chooses the highest and lowest values at a location satisfying the R-H condition

$$x = \frac{\tau}{r^2+1} + r \qquad u = \frac{1}{r^2+1}. \tag{4.10}$$

For any given value of x, we obtain

$$r^3 - x\,r^2 + r + (\tau - x) = 0. \tag{4.11}$$

In the multi-valued zone, this cubic equation will have three real roots, which, when substituted into the second Eq. (4.10), provide three values of u (i.e., the three intersections of the vertical line through x with the wave profile). The Rankine–Hugoniot condition requires the average between the highest and the lowest of these values to determine the slope of the shock curve dividing the two regions where the solution is smooth.

It is interesting to remark that the foregoing picture is an example of a mathematical *catastrophe*, namely, a case in which a perfectly smooth manifold (the parametric surface provided by the method of characteristics, as given by Eq. (3.61)) has singularities in terms of its projection on a plane (because it turns around in such a way that there are vertical tangent planes). The theory of catastrophes, as developed among others by the French mathematician René Thom, was very popular some years ago to explain almost everything under the sun, from the crash of the stock market to the behaviour of a threatened dog.[5] A plot of our equations

$$t = s \qquad x = \frac{s}{r^2+1} + r \qquad u = \frac{1}{r^2+1} \tag{4.12}$$

is shown in Fig. 4.4. Note that this graph is identical to the one in Fig. 3.11 produced numerically by the code of Box 3.3.

For the sake of the illustration, we solve our problem numerically. We remark, however, that the domain of the Rankine–Hugoniot differential equation has a bound-

[5]See [1, 4].

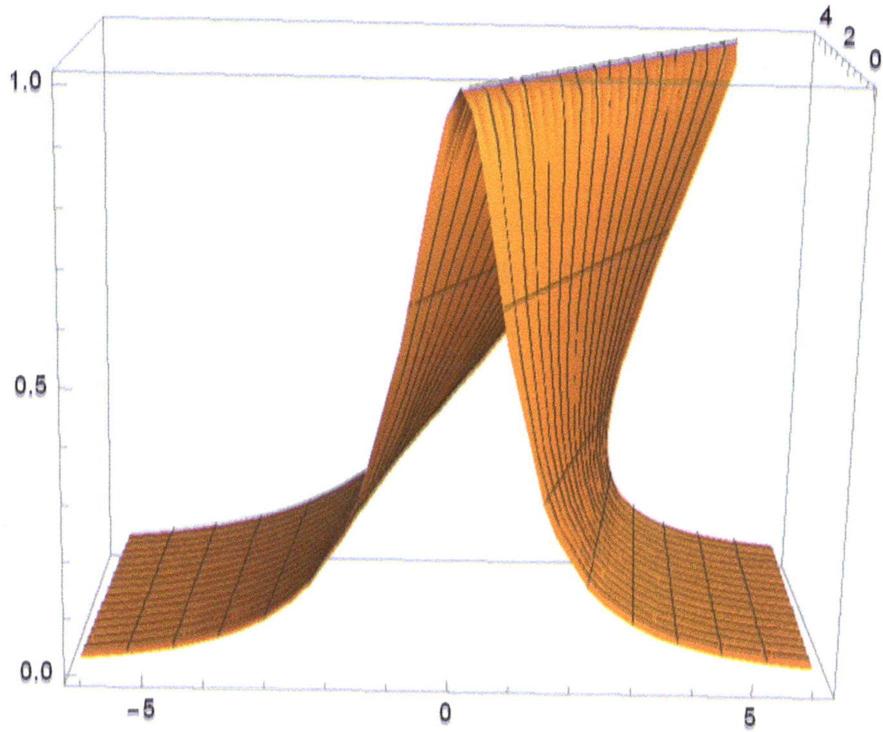

Fig. 4.4 Plot of the multi-valued (parametric) solution surface

ary containing a cusp at the initial point. Thus, the theorem of uniqueness and its numerical implications may not be readily enforceable. Given a cubic equation such as (4.11), one of the roots is always real (since complex roots appear in conjugate pairs) and, when the other two roots happen to be real (in other words, in the multi-valued zone of interest), we need to select the two roots corresponding to the highest and lowest values of u. Choosing these two roots, we solve and plot the solution, as shown in Figs. 4.5 and 4.6.

Figure 4.7 shows the shock solution as a 'chopped' version of the multi-valued solution obtained before. The curved vertical wall projects on the shock curve depicted in Fig. 4.6.

```
(* Roots by Cardano-Tartaglia *)

rr1=x[t]/3-(2^(1/3) (3-x[t]^2))/(3 (-27 t+18 x[t]+2 x[t]^3+3 Sqrt[3] Sqrt[4+27 t^2-36 t x[t]+8 x[t]^2-4 t x[t]^3+4 x[t]^4])^(1/3))+(-27 t+18 x[t]+2 x[t]^3+3 Sqrt[3] Sqrt[4+27 t^2-36 t x[t]+8 x[t]^2-4 t x[t]^3+4 x[t]^4])^(1/3)/(3 2^(1/3))

rr2=x[t]/3+((1+I Sqrt[3]) (3-x[t]^2))/(3 2^(2/3) (-27 t+18 x[t]+2 x[t]^3+3 Sqrt[3] Sqrt[4+27 t^2-36 t x[t]+8 x[t]^2-4 t x[t]^3+4 x[t]^4])^(1/3))-((1-I Sqrt[3]) (-27 t+18 x[t]+2 x[t]^3+3 Sqrt[3] Sqrt[4+27 t^2-36 t x[t]+8 x[t]^2-4 t x[t]^3+4 x[t]^4])^(1/3))/(6 2^(1/3))

rr3=x[t]/3+((1-I Sqrt[3]) (3-x[t]^2))/(3 2^(2/3) (-27 t+18 x[t]+2 x[t]^3+3 Sqrt[3] Sqrt[4+27 t^2-36 t x[t]+8 x[t]^2-4 t x[t]^3+4 x[t]^4])^(1/3))-((1+I Sqrt[3]) (-27 t+18 x[t]+2 x[t]^3+3 Sqrt[3] Sqrt[4+27 t^2-36 t x[t]+8 x[t]^2-4 t x[t]^3+4 x[t]^4])^(1/3))/(6 2^(1/3))

(* Choose highest and lowest *)

maxr=Max[Abs[rr1],Abs[rr2],Abs[rr3]]

minr=Min[Abs[rr1],Abs[rr2],Abs[rr3]]

u1=1/(maxr^2+1)

u2=1/(minr^2+1)

tb=8*Sqrt[3]/9

xbb=Sqrt[3]


(* Solve Rankine-Hugoniot differential equation *)

NDSolve[{x'[t]==0.5*(u1+u2),x[tb]==xbb},x[t],{t,tb,6*tb}]


plot1=ParametricPlot[{{Evaluate[x[t]/.%],t},{(r^2+1)/(2*r)+r,(r^2+1)^2/(2*r)}},{t,tb,6*tb},{r,0.001,3},
plotRange->{{0,5},{0,5}},AxesLabel->{x,t}, AspectRatio->0.75]
```

Fig. 4.5 Mathematica code to solve the Rankine–Hugoniot ODE

4.4 Discontinuous Initial Conditions

4.4.1 Shock Waves

The merit of the example solved in the previous section is that it clearly shows how a shock wave (an essentially discontinuous phenomenon) can develop in a finite time out of *perfectly smooth* initial conditions. If, on the other hand, the initial conditions are themselves discontinuous we may obtain the immediate formation and subsequent propagation of a shock wave. A different phenomenon may also occur, as we will discuss in the next section. Discontinuous initial conditions occur as an idealized

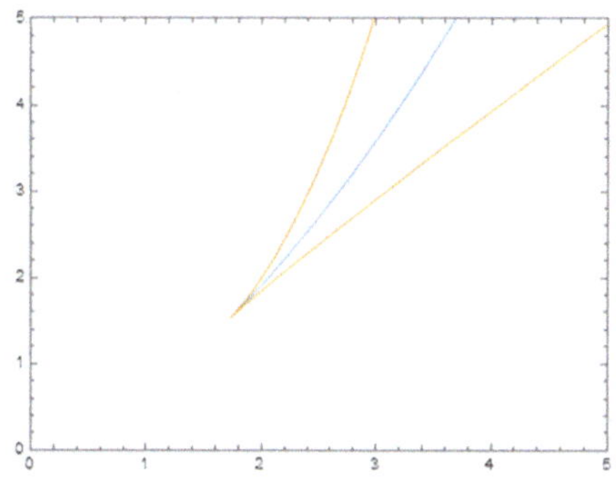

Fig. 4.6 The shock front as a solution of the Rankine–Hugoniot ODE

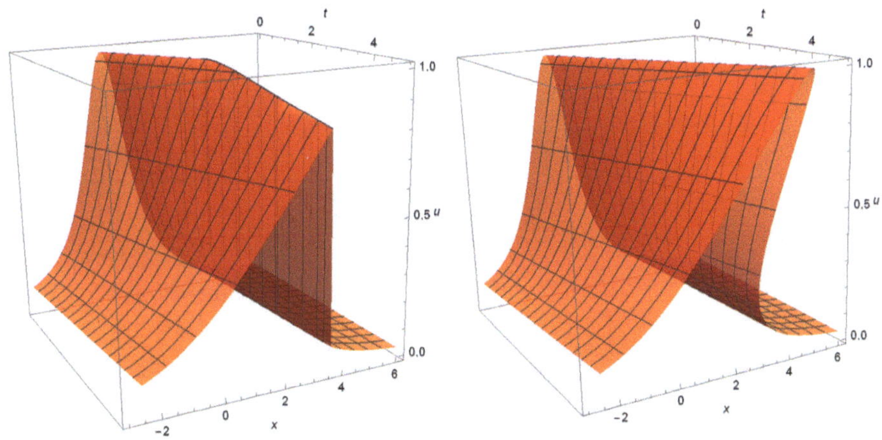

Fig. 4.7 The shock solution (*left*) as a chopped version of the multi-valued solution (*right*)

limit of a steep initial profile, such as that corresponding to the sudden opening of a valve.

Consider again the inviscid Burgers equation (3.50), but with the discontinuous initial conditions

$$u(x,0) = \begin{cases} u^- \text{ for } x \le 0 \\ u^+ \text{ for } x > 0 \end{cases} \tag{4.13}$$

where u^- and u^+ are constants with $u^- > u^+$. These are known as *Riemann* initial conditions. The projected characteristics for this quasi-linear problem are depicted in Fig. 4.8.

We observe that for all $t > 0$ there is a region of intersection of two characteristics, shown shaded in Fig. 4.8. Recalling that for this equation each characteristic carries the constant value u^- or u^+, we conclude that a shock solution is called for in that

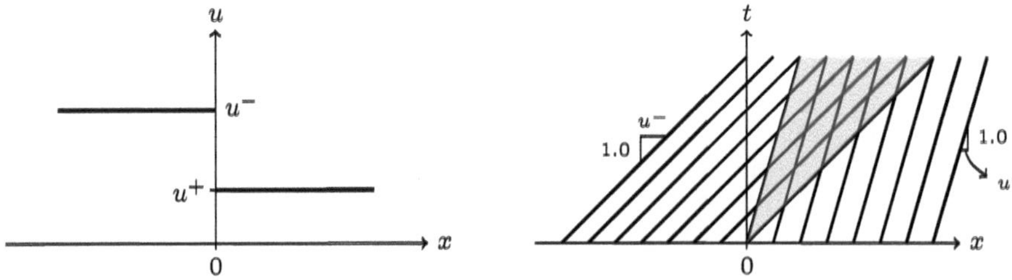

Fig. 4.8 Projected characteristics (*right*) for Riemann conditions (*left*) with $u^- > u^+$

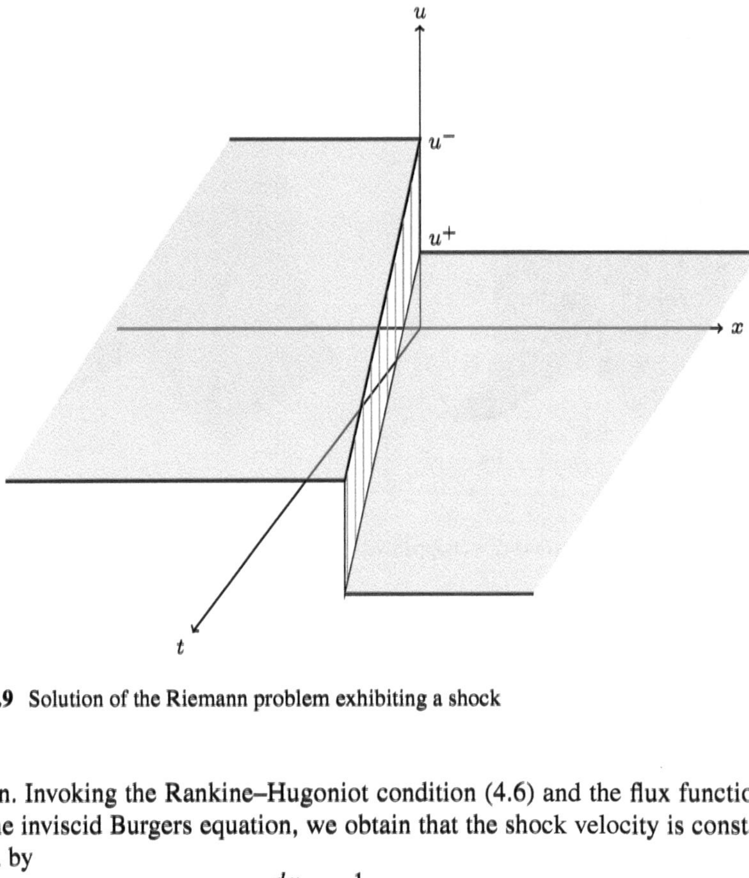

Fig. 4.9 Solution of the Riemann problem exhibiting a shock

region. Invoking the Rankine–Hugoniot condition (4.6) and the flux function (4.8) for the inviscid Burgers equation, we obtain that the shock velocity is constant and given by

$$\frac{dx_s}{dt} = \frac{1}{2}\left(u^- + u^+\right). \tag{4.14}$$

The complete solution is schematically represented in Fig. 4.9

4.4.2 Rarefaction Waves

Even in the case of linear PDEs it is possible to have a situation whereby the (projected) characteristics intersecting the initial manifold do not cover the whole plane. An example is the equation $yu_x + xu_y = u$, for which the characteristics are equilateral hyperbolas. In the case of quasi-linear equations, the projected characteristics depend, of course, on the values of u on the initial manifold, so that the situation just described may be determined by the initial values of the unknown function. Consider again the inviscid Burgers equation with the following Riemann initial conditions

$$u(x, 0) = \begin{cases} u^- & \text{for } x \leq 0 \\ u^+ & \text{for } x > 0 \end{cases} \tag{4.15}$$

where u^- and u^+ are constants with $u^- < u^+$. This is the mirror image of the problem just solved, where the fact that u^- was larger than u^+ led to the immediate formation a shock. The counterpart of Fig. 4.8 is shown in Fig. 4.10. The shaded region is not covered by any characteristic emanating from the initial manifold $t = 0$. Since in physical reality the function u does attain certain values in the shaded region, we need to extend the solution. A possible way to do so is to postulate a constant value of the solution in that region. This device, however, would introduce in general two shocks. It can be verified that no value of the constant would satisfy the Rankine–Hugoniot conditions.[6] A clue as to what needs to be done can be gathered by imagining that the jump in the initial conditions has been replaced by a smooth, albeit steep, transition. Correspondingly, the projected characteristics would now cover the whole half-plane $t > 0$ and would gradually join the existing characteristics in a fan-like manner. Moreover, on each of the new characteristic lines the value of u is constant. In the limit, as the steepness of the transition tends to infinity, the fan of characteristics emanates from the origin, as suggested in Fig. 4.11. To determine the value of u on each of these new characteristic lines, we start by noticing that the function u on the shaded region would have to be of the form $u = f(x/t)$, so as to preserve the constancy on each characteristic in the fan. Introducing the function f into the PDE,

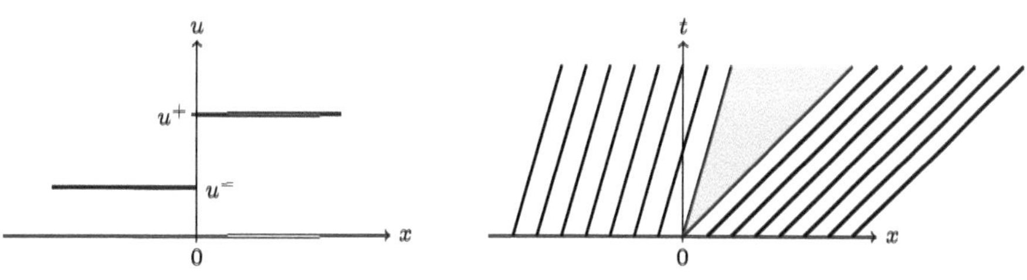

Fig. 4.10 Projected characteristics (*right*) for Riemann conditions (*left*) with $u^- < u^+$

[6]For the possibility of extending the values u^- and u^+ into the shaded region and producing a legitimate shock, see Box 4.1.

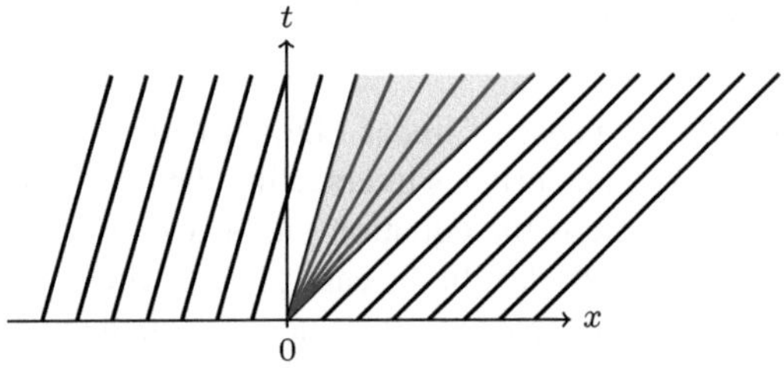

Fig. 4.11 The characteristic fan

we obtain

$$0 = u_t - uu_x = -f'\frac{x}{t^2} + f\frac{f'}{t} = -\frac{f'}{t}\left(\frac{x}{t} - f\right), \tag{4.16}$$

where primes indicate derivatives of f with respect to its argument. Since we have already discarded the constant solutions ($f' = 0$), we are left with

$$u = \frac{x}{t}. \tag{4.17}$$

In other words, for each time $t > 0$ the new solution provides a linear interpolation between the values u^- and u^+ as shown in Fig. 4.12. This solution is continuous and differentiable everywhere except on the two extreme characteristics. It is called a *rarefaction wave*, since it corresponds to a softening of the initial conditions as time goes on. In applications to gas dynamics, a rarefaction wave is a wave of decompression.

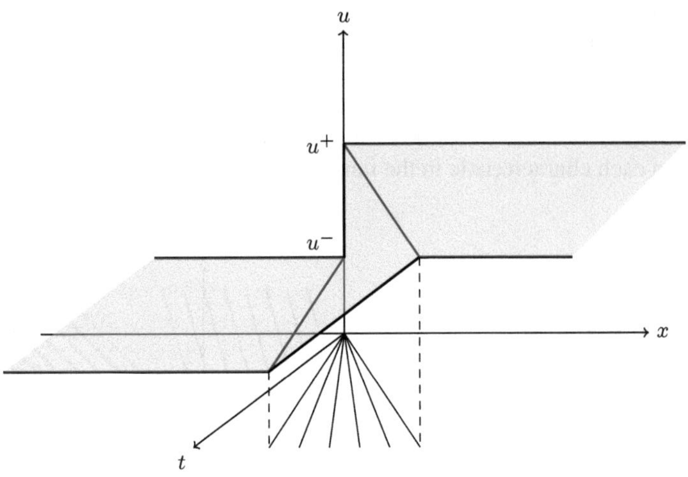

Fig. 4.12 Solution containing a rarefaction wave

Box 4.1 The entropy condition
We have discarded the imposition of a constant solution within the shaded area of Fig. 4.10 in favour of the linearly interpolated condition on the grounds that the accompanying shocks would have violated the Rankine-Hugoniot condition. It is possible, however, to propose other solutions containing a shock that do not violate the Rankine-Hugoniot condition (see Exercise 4.3). In particular, it is possible to extend the solutions u^- and u^+ into the shaded area and have them meet at a shock line that satisfies the Rankine Hugoniot condition. The continuous solution is to be preferred over the solutions of this kind because it satisfies the following, physically based, *entropy condition* [3]: "The characteristics starting [from the initial manifold $t = 0$] on either side of the discontinuity curve when continued in the direction of increasing t intersect the line of discontinuity." Or: "Every point can be connected by a backward drawn characteristic to a point in the initial line."

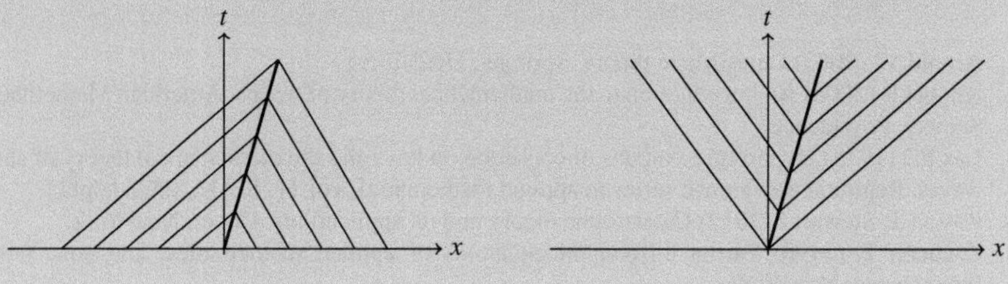

"Good" shock "Bad" shock

Exercises

Exercise 4.1 ([2], *p. 153*) Write the Rankine–Hugoniot condition for the traffic equation (2.30). Assume that some cars have already stopped at a red traffic light and that they are packed at the maximum density. Assume, moreover, that the cars approaching the end of the stopped line are traveling at a constant speed $v_0 < v_{max}$. Find the speed at which the tail of the stopped traffic backs up as more and more cars join in.

Exercise 4.2 The Burgers equation for a dust: Imagine a one-dimensional flow of non-interacting particles with no external forces. Show that this phenomenon is described exactly by the inviscid Burgers equation (3.50). [Hint: interpret $u(x, t)$ as the velocity field expressed in a fixed inertial coordinate system and notice that the particle velocities remain constant.] Notice that if all the particles are moving, say, to the right and if the initial conditions are such that the velocity profile increases to the right, then there is no danger that any particles will catch up with other particles. On the other hand, if any part of the initial velocity profile has a decreasing pattern, some particles will eventually catch up with those to their right and a 'snowballing effect' (that is, a shock) will occur. The Rankine–Hugoniot condition is more difficult to interpret intuitively, but it may be useful to try. [Hint: imagine identical particles equally spaced moving to the right at constant speed and encountering identical particles at rest. Assume perfectly plastic multiple collisions and use induction.]

Exercise 4.3 Show that for the initial conditions (4.15) a mathematical alternative to the rarefaction wave is the following shock solution

$$u = \begin{cases} u^- \text{ for } x \le \frac{1}{2}\left(u^- + u^+\right) \\ u^+ \text{ for } x > \frac{1}{2}\left(u^- + u^+\right) \end{cases}$$

Specifically, verify that this solution satisfies the Rankine–Hugoniot condition. Does it satisfy the entropy condition?

References

1. Arnold VI (2003) Catastrophe theory. Springer, Heidelberg
2. Knobel R (2000) An introduction to the mathematical theory of waves. American Mathematical Society, Providence
3. Lax PD (1973) Hyperbolic systems of conservation laws and the mathematical theory of shock waves. Regional conference series in applied mathematics, vol 11. SIAM, Philadelphia
4. Poston T, Stewart I (2012) Catastrophe theory and its applications. Dover, New York
5. Zauderer E (1998) Partial differential equations of applied mathematics, 2nd edn. Wiley-Interscience, New York

Chapter 5
The Genuinely Nonlinear First-Order Equation

The general non-linear first-order partial differential equation requires a deeper analysis than its linear and quasi-linear counterparts. Instead of a field of characteristic directions, the non-linear equation delivers a one-parameter family of directions at each point of the space of dependent and independent variables. These directions subtend a local cone-like surface known as the Monge cone. Augmenting the underlying space by adding coordinates representing the first partial derivatives of the unknown field, however, it is possible to recover most of the features of the quasi-linear case so that, ultimately, even the solution of the general non-linear equation can be reduced to the integration of a system of ordinary differential equations.

5.1 Introduction

From the treatment of the previous chapters, it is quite clear that quasi-linear equations can be characterized geometrically in a manner not very different from that of linear equations. It is true that the behaviour of quasi-linear equations is richer in content due to the fact that projected characteristics may intersect and thus give rise to the appearance of shocks. Nevertheless, the basic interpretation of the first-order PDE as a field of directions and the picture of a solution as a surface fitting this field are the same, whether the equation happens to be linear or quasi linear. In a genuinely nonlinear first-order PDE, on the other hand, these basic geometric ideas are lost and must be replaced by somewhat more general counterparts. Remarkably, in spite of the initially intimidating nature of the problem, the nature of the final result is analogous to that of the linear and quasi linear cases. Namely, the construction of a solution of a genuinely nonlinear PDE turns out to be based entirely upon the solution of a system of ODEs.

Nonlinear first-order PDEs appear in a variety of applications, most notably in the characterization of wave fronts arising from a system of linear second-order

M. Epstein, *Partial Differential Equations*, Mathematical Engineering,
DOI 10.1007/978-3-319-55212-5_5

PDEs describing wave phenomena. Another important application of the theory is encountered in the Hamilton–Jacobi formulation and its applications to Mechanics. In this chapter, we will start with an exposition of the basic concepts pertaining to the analysis of a single nonlinear first-order PDE in two independent variables and then extend the analysis to more general situations.

5.2 The Monge Cone Field

The general first-order PDE in two variables is of the form

$$F(x, y, u, u_x, u_y) = 0. \tag{5.1}$$

Here, F is a differentiable function of its 5 arguments. We seek a solution, that is, a function $u = u(x, y)$ satisfying Eq. (5.1). Such a solution can be regarded as a surface in the three-dimensional space of (Cartesian) coordinates x, y, u, just as in the quasi linear case. Assuming this solution to be differentiable (as we must, unless we are talking about generalized solutions), this surface will have a well-defined tangent plane at each point of its domain. The equation of the tangent plane at a generic point (x_0, y_0, u_0) lying on this surface is given by

$$u - u_0 = u_x(x - x_0) + u_y(y - y_0). \tag{5.2}$$

It is understood that in this equation the derivatives u_x and u_y are evaluated at the point (x_0, y_0) and that $u_0 = u(x_0, y_0)$, since the point of tangency must belong to the surface. The equation of the tangent plane we have just written is a direct consequence of the very definition of partial derivatives of a function of two variables. The vector with components $\{u_x, u_y, -1\}$ is perpendicular to the surface at the point of tangency.

What our PDE (5.1) tells us is that the two slopes of the tangent plane of any putative solution surface are not independent of each other, but are rather interrelated by the point-wise algebraic condition (5.1), a fact which, in principle, allows us to obtain one slope when the other one is given.

To get a pictorial idea of what this linkage between the two slopes means, let us revisit the linear or quasi linear case, namely,

$$a(x_0, y_0, u_0)\, u_x + b(x_0, y_0, u_0)\, u_y = c(x_0, y_0, u_0). \tag{5.3}$$

Since the normal vector to any solution surface at a point has components proportional to $\{u_x, u_y, -1\}$, we conclude, from Eq. (5.3), that these normal vectors are necessarily perpendicular to a fixed direction in space, namely the (characteristic) direction $\{a, b, c\}$. What this means is that the tangent planes to all possible solution surfaces (at a given point) intersect at the line defined by the characteristic direction, thus forming a *pencil of planes*, as shown in Fig. 5.1. A pencil of planes resembles the pages of a widely open book as they meet at the spine.

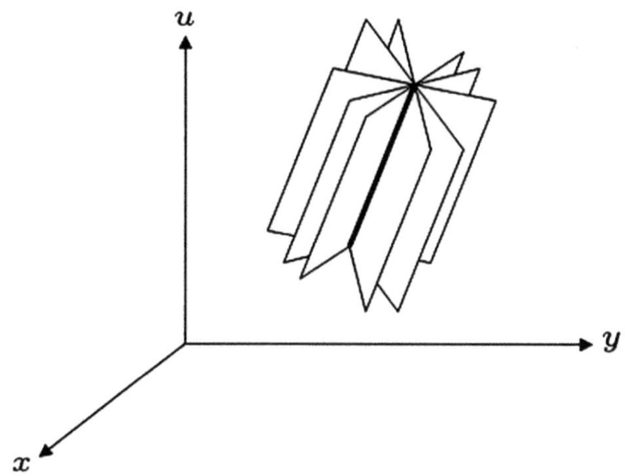

Fig. 5.1 A pencil of planes

In the genuinely non-linear case, on the other hand, no preferred direction $\{a, b, c\}$ is prescribed by the PDE, resulting in the fact that the possible tangent planes (which clearly constitute a one-parameter family of planes at each point of space) do not necessarily share a common line. In general, therefore, we may say that, instead of constituting a pencil of planes around a given line, they envelop a cone-like surface

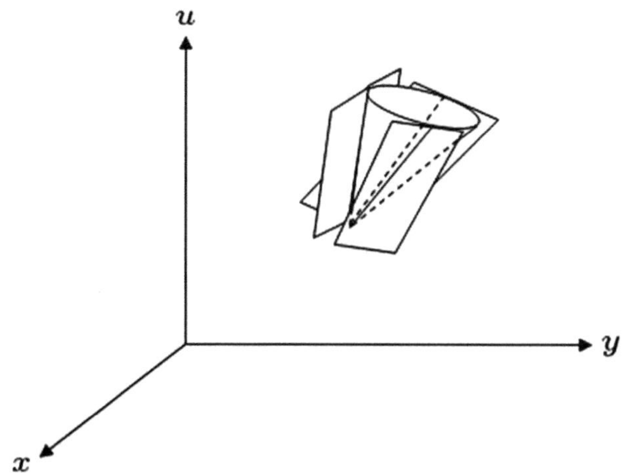

Fig. 5.2 A Monge cone as the envelope of a family of planes at a point

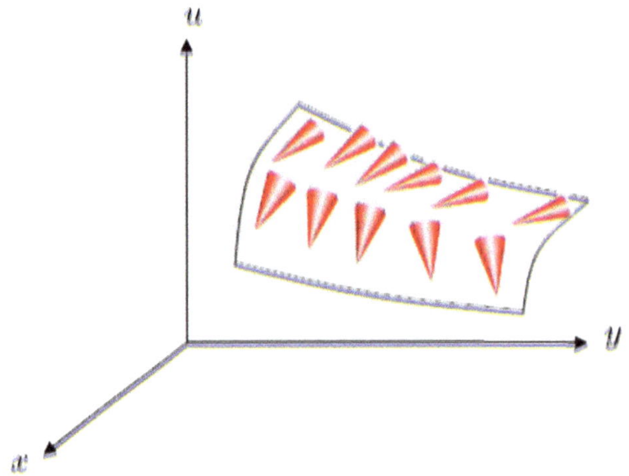

Fig. 5.3 The solution is a surface tangent to the Monge cone field

known as the Monge cone[1] at the given point in space, as shown schematically in Fig. 5.2.[2]

The task of finding a solution to the PDE (5.1) can be regarded geometrically as that of constructing a surface fitting the Monge cone field, in the sense that the surface is everywhere tangent to the local cone, as shown schematically in Fig. 5.3.

5.3 The Characteristic Directions

The Monge cone can be seen at each point as defining not just one characteristic direction (as was the case in the quasi-linear equation) but a one-parameter family of characteristic directions, namely, the family of generators of the cone. To simplify the notation, let us put

$$p = u_x \qquad q = u_y. \tag{5.4}$$

Thus, the PDE (5.1) can be seen at a given point x_0, y_0, u_0 as imposing an algebraic relation between the possible values of the slopes p and q, viz.,

$$F(x_0, y_0, u_0, p, q) = 0. \tag{5.5}$$

[1]In honour of Gaspard Monge (1764–1818), the great French mathematician and engineer, who made seminal contributions to many fields (descriptive geometry, differential geometry, partial differential equations).

[2]There are some mathematical subtleties. For example, we are tacitly assuming that the partial derivatives of the function F with respect to the arguments u_x and u_y do not vanish simultaneously. Also, we are considering a small range of tangent planes, where one of the slopes is a single-valued function of the other.

For each value of one of the slopes, this equation provides us with one[3] value of the other slope. In other words, Eq. (5.5) can be regarded as providing a one-parameter family of slopes, namely,

$$p = p(\alpha) \qquad q = q(\alpha). \tag{5.6}$$

Each value of the parameter α corresponds to a plane. As α varies, these planes envelop the Monge cone. A generator of this cone can loosely be regarded as the intersection of two "neighbouring" planes. According to Eq. (5.2), the equation of the one-parameter family of planes tangent to the cone is explicitly given by

$$u = u_0 = p(\alpha)\,(x - x_0) + q(\alpha)\,(y - y_0). \tag{5.7}$$

To find the intersection between two planes, we need to take the cross product of their normals.[4] The intersection between two neighbouring tangent planes is, therefore, aligned with the vector

$$\mathbf{v} = \begin{vmatrix} \mathbf{i} & \mathbf{j} & \mathbf{k} \\ p & q & -1 \\ p + p'd\alpha & q + q'd\alpha & -1 \end{vmatrix} \tag{5.8}$$

We have indicated with primes the derivatives with respect to the parameter and we have used otherwise standard notation for unit vectors along the coordinate axes. On the other hand, taking the derivative of Eq. (5.5) with respect to the parameter, we obtain

$$F_p\, p' + F_q\, q' = 0, \tag{5.9}$$

where we have used the index notation for partial derivatives of the function F with respect to the subscripted variable. Combining the last two results, we conclude that the direction of the cone generator located in the plane corresponding to the value α of the parameter is given by the vector with components

$$\begin{Bmatrix} F_p \\ \\ F_q \\ \\ p\,F_p + q\,F_q \end{Bmatrix} \tag{5.10}$$

[3]See previous footnote.

[4]The Monge cone is a particular case of an *envelope* of surfaces. In Box 5.1 we present a more general derivation.

Box 5.1 Envelopes of families of surfaces
A one-parameter family of surfaces is given by an equation

$$u = f(x, y, \alpha),$$

where α is the parameter. If the dependence of f on α is smooth, there exists in general an *enveloping surface* $u = U(x, y)$ with the property of being tangent at each of its points to one of the members of the family. This envelope can be obtained by solving the algebraic system

$$u = f(x, y, \alpha) \qquad 0 = f_\alpha(x, y, \alpha),$$

where we are using the subscript notation for partial derivatives. If $f_{\alpha\alpha} \neq 0$ at a value $\alpha = \alpha_0$, the inverse function theorem ensures that we can read off from the second equation $\alpha = g(x, y)$ in an interval around α_0. Plugging this function back in the first equation we obtain

$$u = f(x, y, g(x, y)) = U(x, y).$$

Since $f_\alpha = 0$, we obtain $U_x = f_x + f_\alpha g_x = f_x$, which is the desired property for the envelope.

Intuitively, if we start from the surface $u = f(x, y, \alpha_0)$ and consider the nearby surface $u = f(x, y, \alpha_0 + d\alpha) = f(x, y, \alpha_0) + f_\alpha(x, y, \alpha_0)d\alpha$, these two surfaces intersect along a curve obtained by equating $f(x, y, \alpha_0) = f(x, y, \alpha_0) + f_\alpha(x, y, \alpha_0)d\alpha$. Hence, we obtain the condition $f_\alpha(x, y, \alpha_0) = 0$. So, the system of equations $u = f(x, y, \alpha)$ and $f_\alpha(x, y, \alpha_0) = 0$ can be regarded as a one-parameter family of curves obtained as intersections between neighbouring surfaces of the original family.

At each point of the three-dimensional space with coordinates (x, y, u) we have thus defined a one-parameter family of *characteristic directions*. It is important to observe that in the quasi-linear case, where $F_p = a(x, y, u)$ and $F_q = b(x, y, u)$, expression (5.10) provides us with an explicit system of equations for the characteristic curves, namely, the familiar

$$\frac{dx}{ds} = a(x, y, u) \qquad \frac{dy}{ds} = b(x, y, u) \qquad \frac{du}{ds} = c(x, y, u). \tag{5.11}$$

In the present, genuinely non-linear, case, the system of ODEs

$$\frac{dx}{ds} = F_p \qquad \frac{dy}{ds} = F_q \qquad \frac{du}{ds} = p\ F_p + q\ F_q, \tag{5.12}$$

is under-determined, because the right-hand sides of these equations depend on the 5 variables x, y, u, p, q, which are interrelated by just one extra condition (the PDE itself). But consider now an integral surface $u = u(x, y)$ of our PDE. On this surface, the functions F_p and F_q can be calculated as explicit functions of x and y, and so can p and q. We conclude that, given an integral surface, Eq. (5.12) give us a well-defined vector field attached to this surface. It is not difficult to see that these vectors are actually tangent to the surface. Indeed, the vector with components $\{p, q, -1\}$ is perpendicular to the surface and can be checked to be also perpendicular to the vector defined by the right-hand-sides of Eq. (5.12). We conclude that on any given integral surface of the PDE there exists a well-defined family of *characteristics*, namely, of curves that have everywhere a characteristic direction (or, in other words, the integral surface chooses at each point one of the local Monge cone generators). To write explicitly the system of (five) ODEs satisfied by these characteristic curves associated with a given solution surface, we need to calculate the derivatives of p and q with respect to the curve parameter s. To this effect, we note that, since the PDE (5.1) is satisfied identically, we must also have (by taking derivatives with respect to x and y, respectively)

$$F_x + F_u\ p + F_p\ p_x + F_q\ q_x = 0, \tag{5.13}$$

$$F_y + F_u\ q + F_p\ p_y + F_q\ q_y = 0. \tag{5.14}$$

Combining these results with Eq. (5.12), we can write

$$F_x + F_u\ p + \frac{dx}{ds}\ p_x + \frac{dy}{ds}\ q_x = 0, \tag{5.15}$$

$$F_y + F_u\ q + \frac{dx}{ds}\ p_y + \frac{dy}{ds}\ q_y = 0. \tag{5.16}$$

But, since 'mixed partials are equal', we have that $p_y = q_x$, so that ultimately Eqs. (5.15) and (5.16) can be written as

$$\frac{dp}{ds} = -F_x - F_u\ p, \tag{5.17}$$

$$\frac{dq}{ds} = -F_y - F_u\ q. \tag{5.18}$$

Equations (5.12), (5.17) and (5.18) constitute a system of five first-order ODEs satisfied by the characteristic curves contained in a given solution surface. Suppose now, vice versa, that these five equations had been given a priori, without any knowledge of any particular solution surface. This system 'happens to have' the function F as a first integral. What this means is that this function attains a constant value on every integral curve of the given system of ODEs. Indeed, we check that

$$\frac{dF}{ds} = F_x \frac{dx}{ds} + F_y \frac{dy}{ds} + F_u \frac{du}{ds} + F_p \frac{dp}{ds} + F_q \frac{dq}{ds} = 0, \tag{5.19}$$

where we have used Eqs. (5.12), (5.17) and (5.18). If we now single out of all the possible solutions of this system of ODEs those for which $F = 0$, we obtain a special (three-parameter) sub-family of solutions called *characteristic strips* of the PDE. The reason for this terminology is that each such solution can be seen as an ordinary curve, $x(s), y(s), u(s)$, each point of which carries a plane element, that is, the two slopes $p(s), q(s)$. The image to have in mind is that of a tapeworm. If a characteristic strip has an element x, y, u, p, q in common with a solution surface, the whole strip must belong to this surface.

5.4 Recapitulation

It may not be a bad idea to review the basic geometric ideas we have been working with.[5] A *point* is a triple (x_0, y_0, u_0). A *plane element* is a quintuple (x_0, y_0, u_0, p, q). Thus, a plane element consists of a point and two slopes defining a plane passing through that point, namely,

$$u - u_0 = p\ (x - x_0) + q\ (y - y_0). \tag{5.20}$$

The vector with components $\{p, q, -1\}$ is perpendicular to this plane. A smooth one-parameter family of plane elements $(x(s), y(s), u(s), p(s), q(s))$ consists of a smooth curve at each point of which there is a plane attached. These planes may or may not be tangential to the underlying curve, also called the *support curve* $(x(s), y(s), u(s))$. If they are, we say that $(x(s), y(s), u(s), p(s), q(s))$ is a *strip*. Notice that so far all these concepts are completely independent of any PDE; they are purely geometric definitions. We now ask the question: what is the condition for a one-parameter quintuple to constitute a strip? The answer is quite straightforward once we observe that for this to be the case the normal to the plane attached at a point must be perpendicular to the support curve at that point. In other words, we need to satisfy

$$p\ \frac{dx}{ds} + q\ \frac{dy}{ds} - 1\ \frac{du}{ds} = 0. \tag{5.21}$$

This equation is appropriately known as the *strip condition*. The above geometric concepts are illustrated in Fig. 5.4.

Having established these basic definitions, let us introduce the first-order non-linear PDE

$$F(x, y, u, p, q) = 0, \tag{5.22}$$

[5] We follow the terminology of [3].

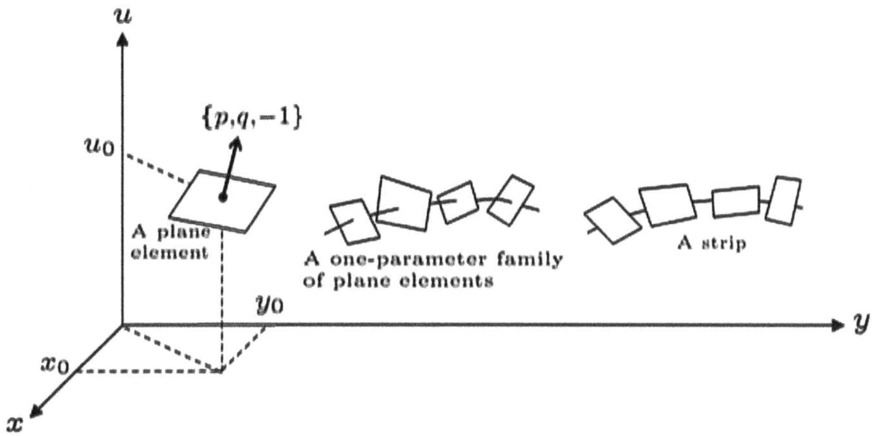

Fig. 5.4 Planes and strips

where we are using the notation (5.4). We form the following system of five ODEs, which we call the *characteristic system* associated with the given PDE:

$$\frac{dx}{ds} = F_p \qquad \frac{dy}{ds} = F_q \qquad \frac{du}{ds} = p\ F_p + q\ F_q$$
$$\frac{dp}{ds} = -F_x - F_u\ p \qquad \frac{dq}{ds} = -F_y - F_u\ q. \tag{5.23}$$

Any solution (integral curve) of this system can be obviously viewed as a one-parameter family of plane elements supported by a curve. We claim that this one-parameter family is necessarily a strip, which will be called a *characteristic strip*. The proof of this assertion follows directly from substitution of the first three equations of the system (5.23) into the strip condition (5.21).

To pin down a characteristic strip (it being the solution of a system of ODEs), we only need to specify any (initial) plane element belonging to it. We have shown that the function F defining our PDE is a first integral of its characteristic system. Therefore, if any one plane element of a characteristic strip satisfies the equation $F = 0$, so will the whole characteristic strip to which it belongs. This result can be interpreted as follows. If a plane element is constructed out of a point on an integral surface of the PDE and of the tangent plane to this surface at that point, then the strip that this element uniquely determines has a support curve that belongs to the integral surface (that is, a characteristic curve) and the plane elements of this strip are made up of the tangent planes to the integral surface at the corresponding points. Two characteristic strips with $F = 0$ whose support curves have a common point with a common tangent, must coincide. Therefore, two integral surfaces having a common point and a common tangent plane thereat, must have a whole characteristic strip in common (that is, they share the whole support curve and are tangential to each other along this curve).

5.5 The Cauchy Problem

The Cauchy (or initial) problem for a nonlinear PDE is essentially the same as for its linear or quasi-linear counterpart. Given an initial curve in the plane

$$x = \hat{x}(r) \qquad y = \hat{y}(r), \tag{5.24}$$

on which the values of the unknown function have been prescribed as

$$u = \hat{u}(r), \tag{5.25}$$

the Cauchy problem deals with the possibility of finding a solution of the PDE over a neighbourhood of the initial curve in such a way that it attains the prescribed values over this curve. Equations (5.24) and (5.25) constitute the parametric equations of a curve in three-dimensional space. The Cauchy problem can, therefore, be rephrased as follows: to find an integral surface of the PDE containing this space curve.

In the case of linear and quasi-linear equations, the solution to this problem was based on the construction of the one-parameter family of characteristics issuing from the various points of this space curve. The situation in a genuinely non-linear first-order PDE is more delicate, since what we have at our disposal is not a collection of characteristic curves, but rather of characteristic strips. The task of constructing the solution must start therefore by extending in a unique way the initial data to a (non-characteristic) strip, and only then solving the differential equations of the characteristic strips to generate a one-parameter family. We will need to show how this extension is accomplished and to prove that the one-parameter family of characteristic support curves is indeed an integral surface. These tasks are somewhat more difficult than in the case of the quasi-linear equation, but the fundamental idea of reducing the Cauchy problem of a first-order PDE to the integration of a system of ODEs remains the same.

The (non-characteristic) initial strip supported by the given curve will have a parametric representation consisting of the equations of the supporting curve (5.24), (5.25) and two additional equations

$$p = \hat{p}(r) \qquad q = \hat{q}(r), \tag{5.26}$$

providing the slopes of the tangent plane as functions of the running parameter r. To determine these two functions, we have at our disposal two equations. The first equation is the strip condition (5.21), guaranteeing that each plane element contains the local tangent to the curve, namely,

$$\hat{p}\,\frac{d\hat{x}}{dr} + \hat{q}\,\frac{d\hat{y}}{dr} = \frac{d\hat{u}}{dr}. \tag{5.27}$$

The second equation at our disposal is the PDE itself (which we clearly want to see satisfied on this initial strip), that is,

$$F(\hat{x}(r), \hat{y}(r), \hat{u}(r), \hat{p}(r), \hat{q}(r)) = 0. \tag{5.28}$$

We note that these two equations constitute, at each point, merely algebraic relations between the two unknown quantities $\hat{p}(r), \hat{q}(r)$. To be able to read off these unknowns at a given point in terms of the remaining variables, we need the corresponding Jacobian determinant

$$J = \begin{vmatrix} \frac{d\hat{x}}{dr} & \frac{d\hat{y}}{dr} \\ \\ F_p & F_q \end{vmatrix} \tag{5.29}$$

not to vanish at that point. By continuity, there will then exist a neighbourhood of this point with the same property. In this neighbourhood (which we will assume to be the whole curve) we can obtain the desired result by algebraic means. Using each plane element thus found as an initial condition for the system of characteristic ODEs, and setting the parameter s of the characteristic strips thus obtained to 0 at the point of departure, we obtain a one-parameter family of characteristic strips, namely,

$$x = x(r, s) \qquad y = y(r, s) \qquad u = u(r, s) \qquad p = p(r, s) \qquad q = q(r, s). \tag{5.30}$$

We claim that the first three equations in (5.30) constitute an integral surface of the PDE. It is clear that on the surface represented parametrically by these three equations, the PDE is satisfied as an algebraic relation between the five variables x, y, u, p, q. What remains to be shown is that we can read off the parameters r and s in terms of x and y from the first two equations and that, upon entering these values into the third equation and calculating the partial derivatives u_x and u_y, we recover, respectively, the values of p and q given by the last two equations in (5.30). We will omit the proof of these facts.[6]

5.6 An Example

To illustrate all the steps involved in the solution of a non-linear first-order PDE by the method of characteristic strips, we will presently solve a relatively simple example.[7] The problem consists of finding a solution of the PDE

$$u = u_x^2 - u_y^2, \tag{5.31}$$

[6]See [3].

[7]This example is suggested as an exercise in [4], p. 66.

which on the x-axis attains the value

$$u = -\frac{1}{4}x^2. \tag{5.32}$$

Solution: Our first task is to construct a (hopefully non-characteristic) strip supported by the initial 3-D curve, which can be parametrized as

$$x = r \qquad y = 0 \qquad u = -\frac{1}{4}r^2. \tag{5.33}$$

The strip condition (5.27) yields

$$1\,p + 0\,q = -\frac{1}{2}r. \tag{5.34}$$

Moreover, the PDE itself, i.e. Eq. (5.31), yields

$$-\frac{1}{4}r^2 = p^2 - q^2. \tag{5.35}$$

Solving the system of equations (5.34) and (5.35), we obtain

$$p = -\frac{1}{2}r \qquad q = \pm\frac{\sqrt{2}}{2}r. \tag{5.36}$$

This completes the strip over the support curve (5.33). It is important to notice that, due to the non-linearity of the PDE, we happen to obtain two different possibilities for the initial strip, each of which will give rise to a different solution of the equation.

Our next task is to obtain and solve the characteristic system of ODEs. Writing the PDE (5.31) in the form

$$F(x, y, u, p, q) = u - p^2 + q^2 = 0, \tag{5.37}$$

Equation (5.23) result in the system

$$\frac{dx}{ds} = -2p \qquad \frac{dy}{ds} = 2q \qquad \frac{du}{ds} = -2p^2 + 2q^2 \qquad \frac{dp}{ds} = -p \qquad \frac{dq}{ds} = -q. \tag{5.38}$$

This system is easily integrated to

$$x = 2Ae^{-s} + C, \tag{5.39}$$

$$y = -2Be^{-s} + D, \tag{5.40}$$

$$u = (A^2 - B^2)e^{-2s} + E, \tag{5.41}$$

$$p = Ae^{-s}, \tag{5.42}$$

$$q = Be^{-s}, \tag{5.43}$$

where A, B, C, D, E are adjustable constants of integration. Setting $s = 0$, we make each of these expressions equal to the respective counterpart in Eq. (5.33) or (5.36) and thus obtain

$$A = -\frac{1}{2}r \qquad B = \pm\frac{\sqrt{2}}{2}r \qquad C = 2r \qquad D = \pm\sqrt{2}r \qquad E = 0. \tag{5.44}$$

Introducing these values into Eqs. (5.39)–(5.41), we obtain the following parametric equation of the integral surface:

$$x = r(2 - e^{-s}), \tag{5.45}$$

$$y = \pm\sqrt{2}r(1 - e^{-s}), \tag{5.46}$$

$$u = -\frac{1}{4}r^2 e^{-2s}. \tag{5.47}$$

We now solve Eqs. (5.45) and (5.46) for r and s to obtain

$$r = x \mp \frac{\sqrt{2}}{2}y, \tag{5.48}$$

$$e^{-s} = 2 - \frac{x}{x \mp \sqrt{2}y/2}. \tag{5.49}$$

Substituting these values in Eq. (5.47), we obtain the desired solution as the integral surface

$$u = -\frac{1}{4}\left(\pm\sqrt{2}y - x\right)^2. \tag{5.50}$$

Geometrically, the solution is either one of two horizontal oblique parabolic cylinders. Either cylinder contains the initial data.

5.7 More Than Two Independent Variables

5.7.1 *Quasi-linear Equations*

Our treatment of first-order linear and non-linear PDEs has been constrained so far to the case of two independent variables. The main reason for this restriction has been to enable the visualization of the solutions as surfaces in $\mathbb{R}^3$ and thus to foster geometric reasoning. The generalization to an arbitrary number of independent variables is quite straightforward. Consider the quasi-linear equation

$$a_1(x_1, \ldots, x_m, u)\frac{\partial u}{\partial x_1} + \cdots + a_m(x_1, \ldots, x_m, u)\frac{\partial u}{\partial x_m} = c(x_1, \ldots, x_m, u), \quad (5.51)$$

for a function u of the m variables $x_1, \ldots, x_m$. In a more economic notation this equation can be also written as

$$a_i \; p_i = c, \quad (5.52)$$

where we are using the summation convention in the range $1, \ldots, m$ and the notation introduced in (3.2).

The characteristic vector field associated with this equation is defined as the vector field in $\mathbb{R}^{m+1}$ with components $a_1, \ldots, a_m, c$. The integral curves of this field, namely, the solutions of the system of characteristic ODEs

$$\frac{dx_i}{ds} = a_i \quad (i = 1, \ldots, m) \qquad \frac{du}{ds} = c, \quad (5.53)$$

are the *characteristic curves* of the PDE. As before, we can prove that if a characteristic curve has one point in common with a solution $u = u(x_1, \ldots, x_m)$, then the whole characteristic curve belongs to this solution. In geometric terms, a function $u = u(x_1, \ldots, x_m)$ is a *hyper-surface* of dimension m in $\mathbb{R}^{m+1}$. The Cauchy problem consists of finding a solution when initial data have been given in a hyper-surface Γ of dimension $m - 1$. In parametric form, such a hyper-surface can be represented as

$$x_i = \hat{x}_i(r_1, \ldots, r_{m-1}) \quad (i = 1, \ldots, m) \qquad u = \hat{u}(r_1, \ldots, r_{m-1}), \quad (5.54)$$

where $r_1, \ldots, r_{m-1}$ are parameters. Assuming that this initial hyper-surface is nowhere tangent to a characteristic curve, the problem has a unique solution in a neighbourhood of Γ. This solution can be constructed as the $(m - 1)$-parameter family of characteristics issuing from the points of Γ

Example 5.1 Solve the Burgers-like quasi-linear initial-value problem

$$u_t + uu_x + u^2 u_y = 0 \qquad u(x, y, 0) = x + y, \quad (5.55)$$

for a function $u = u(x, y, t)$ in the half-space $t \geq 0$.

Solution: The characteristic equation are

$$\frac{dx}{ds} = u \qquad \frac{dy}{ds} = u^2 \qquad \frac{dt}{ds} = 1 \qquad \frac{du}{ds} = 0. \quad (5.56)$$

The characteristics (integral curves) are obtained as

$$x = Ds + A \qquad y = D^2 s + B \qquad t = s + C \qquad u = D, \tag{5.57}$$

where A, B, C, D are integration constants. The initial condition can be written in parametric form as

$$x = r_1 \qquad y = r_2 \qquad t = 0 \qquad u = r_1 + r_2, \tag{5.58}$$

where r_1, r_2 are parameters.

Setting $s = 0$ at the initial manifold and enforcing the initial conditions, we obtain

$$A = r_1 \qquad B = r_2 \qquad C = 0 \qquad D = r_1 + r_2. \tag{5.59}$$

The solution hyper-surface in parametric form reads, therefore,

$$x = (r_1 + r_2)s + r_1 \qquad y = (r_1 + r_2)^2 s + r_2 \qquad t = s \qquad u = r_1 + r_2. \tag{5.60}$$

We need to express the parameters in terms of the original independent variable x, y, t. Adding the first two equations and enforcing the third, we obtain

$$x + y = (r_1 + r_2)^2 t + (r_1 + r_2)(t + 1), \tag{5.61}$$

whence

$$r_1 + r_2 = \frac{-(t+1) \pm \sqrt{(t+1)^2 + 4t(x+y)}}{2t}. \tag{5.62}$$

Invoking the fourth parametric equation we can write the final result as

$$u = \frac{-(t+1) + \sqrt{(t+1)^2 + 4t(x+y)}}{2t}. \tag{5.63}$$

The choice of the positive sign has to do with the imposition of the initial condition. Because of the vanishing denominator at $t = 0$ we verify that

$$\lim_{t \to 0} \frac{-(t+1) + \sqrt{(t+1)^2 + 4t(x+y)}}{2t} = \lim_{t \to 0} \frac{-1 + \frac{2(t+1)+4(x+y)}{2\sqrt{(t+1)^2+4t(x+y)}}}{2} = x + y. \tag{5.64}$$

Our solution (5.63) is not defined when the radicand is negative, namely, in the subspace of $\mathbb{R}^3$ defined as

$$(t+1)^2 + 4t(x+y) < 0. \tag{5.65}$$

It is not difficult to verify that at the boundary of this domain the Jacobian determinant $\partial(x, y, t)/\partial(r_1, r_2, t)$ vanishes. Moreover, the t-derivative of the solution at the initial manifold is infinite.

5.7.2 Non-linear Equations

The most general non-linear PDE of the first order has the form

$$F(x_1, \ldots, x_m, u, p_1, \ldots p_m) = 0, \tag{5.66}$$

where the notation of Sect. 3.1 is used, namely, $p_i = u_{,i}$. We define the *characteristic system* of ODEs associated with the PDE (5.66) as

$$\frac{dx_i}{ds} = F_{p_i} \qquad \frac{du}{ds} = p_i F_{p_i} \qquad \frac{dp_i}{ds} = -F_{x_i} - F_u p_i, \tag{5.67}$$

where the summation convention in the range $i = 1, \ldots, m$ is understood. These equations are the m-dimensional analogues of Eqs. (5.12), (5.17) and (5.18).

The function F itself is a first integral of the characteristic system. Indeed, always using the summation convention, on every solution of the characteristic system we obtain

$$\begin{aligned}\frac{dF}{ds} &= F_{x_i}\frac{dx_i}{ds} + F_u\frac{du}{ds} + F_{p_i}\frac{dp_i}{ds} \\ &= F_{x_i}F_{p_i} + F_u\ p_i F_{p_i} + F_{p_i}\left(-F_{x_i} - F_u p_i\right) = 0.\end{aligned} \tag{5.68}$$

A solution of the characteristic system (5.67) on which $F = 0$ is called a *characteristic strip*. Suppose that a characteristic strip has a point in common with a solution $u = u(x_1, \ldots, x_m)$. By this we mean, of course, that the strip coincides with the solution at that point and that they both have the same derivatives $p_1, \ldots, p_m$. Then, since $F = 0$ on the strip, the characteristic strip lies entirely on the solution.

The Cauchy problem for the non-linear first-order PDE stipulates data on an initial manifold of dimension $m - 1$. This initial manifold is given, generally, in terms of some differentiable functions depending on $m - 1$ parameters r_i $(i = 1, \ldots, m - 1)$, that is,

$$x_i = \hat{x}_i(r_1, \ldots, r_{m-1}) \qquad i = 1, \ldots, m, \tag{5.69}$$

on which we specify $u = \hat{u}(r_1, ..., r_{m-1})$. By analogy with the two-dimensional case, we extend the initial data as some functions $p_i = \hat{p}_i(r_1, \ldots r_{m-1})$. To this end, we impose the *strip conditions*

$$\hat{p}_i\,\frac{\partial \hat{x}_i}{\partial r_k} = \hat{u}_{r_k} \qquad k = 1, \ldots, m - 1, \tag{5.70}$$

and the PDE itself evaluated at the initial manifold, that is,

$$F(\hat{x}_1, \ldots, \hat{x}_m, \hat{u}, \hat{p}_1, \ldots \hat{p}_m) = 0. \tag{5.71}$$

Equations (5.70) and (5.71) constitute an algebraic system of m equations that can, in principle, be solved for the values of $p_1, \ldots, p_m$ for each point $r_1, \ldots, r_{m-1}$ on the initial manifold. By the inverse function theorem, this is possible if the Jacobian determinant

$$J = \begin{vmatrix} \frac{\partial \hat{x}_1}{\partial r_1} & \cdots & \frac{\partial \hat{x}_m}{\partial r_1} \\ \cdot & \cdots & \cdot \\ \cdot & \cdots & \cdot \\ \cdot & \cdots & \cdot \\ \frac{\partial \hat{x}_1}{\partial r_{m-1}} & \cdots & \frac{\partial \hat{x}_m}{\partial r_{m-1}} \\ F_{p_1} & \cdots & F_{p_m} \end{vmatrix} \tag{5.72}$$

does not vanish over the initial manifold. If this condition is satisfied, we can build a solution of the PDE that contains the initial data by constructing the $(m-1)$-parameter family of characteristic strips issuing from each point of the initial manifold. This is achieved by means of the by now familiar procedure of setting $s = 0$ in the general expression for the characteristic strips and equating with the corresponding values at each point of the initial manifold (extended as explained above). In this way, the 'constants' of integration are obtained in terms of the parameters $r_1, \ldots, r_{m-1}$. The solution is thus obtained in the parametric form

$$x_i = x_i(s, r_1, \ldots, r_{m-1}) \quad (i = 1, \ldots, m) \qquad u = u(s, r_1, \ldots, r_{m-1}). \tag{5.73}$$

To eliminate the parameters, we need to guarantee that the Jacobian matrix $\partial(x_1, \ldots, x_m)/\partial(s, r_1, \ldots, r_{m-1})$ does not vanish. But this determinant J is precisely given by Eq. (5.72), as shown in Exercise 5.4. Since J does not vanish, by assumption, on the initial manifold (where $s = 0$), by continuity it will not vanish on a neighbourhood of the initial manifold. Therefore, in principle, we obtain a local solution in the desired form $u = u(x_1, \ldots, x_m)$.

5.8 Application to Hamiltonian Systems

5.8.1 Hamiltonian Systems

A *Hamiltonian system* consists of *states* described in terms of n *generalized coordinates* $q_1, \ldots, q_n$ and n *generalized momenta* $p_1, \ldots, p_n$. The physical properties of the system are completely characterized by a single scalar function $H = H(q_1, \ldots, q_n, p_1, \ldots, p_n, t)$, where t is a time-like variable. This function, assumed to be differentiable, is called the *Hamiltonian* of the system. Finally, the *evolution* of the system in time is governed by Hamilton's equations, viz.,

$$\frac{dq_i}{dt} = \frac{\partial H}{dp_i} \qquad \frac{dp_i}{dt} = -\frac{\partial H}{dq_i} \qquad i = 1, \ldots, n. \tag{5.74}$$

Note that these are ODEs. The partial derivatives on the right hand side are known functions of q, p, t obtained by differentiating the Hamiltonian function. A solution of this system constitutes a *trajectory*. A trajectory can be regarded as a curve in $\mathbb{R}^{2n}$, the space with coordinates $q_1, \ldots, q_n, p_1, \ldots, p_n$ known as the *phase space* of the system.

Although Hamilton's equations were originally formulated by starting from Lagrangian Mechanics and effecting a certain Legendre transformation relative to the generalized velocities, Hamiltonian systems can arise independently in Mechanics and in other branches of Physics such as Optics, General Relativity and Quantum Mechanics.

5.8.2 Reduced Form of a First-Order PDE

At first sight, there seems to be no relation whatsoever between Hamiltonian systems (governed by a system of ODEs) and first-order PDEs. It is true that, as we have pointed out repeatedly, the Cauchy problem for a first-order PDE can be reduced to the integration of the system of characteristic ODEs. But is the converse possible or even worthwhile? To work towards an answer to these questions, let us start by considering a special kind of first-order PDE, which we call *reduced*. This a PDE of the form

$$F(x_1, \ldots, x_n, p_1, \ldots p_n) = 0. \tag{5.75}$$

The peculiarity of this PDE is that the unknown function u *does not appear explicitly* in the function F. The characteristic strips for this equation are given, according to (5.67), by

$$\frac{dx_i}{ds} = F_{p_i} \qquad \frac{dp_i}{ds} = -F_{x_i} \qquad \frac{du}{ds} = p_i F_{p_i} \qquad i = 1, \ldots, n, \tag{5.76}$$

with the summation convention implied in the range $1, \ldots, n$.

If we compare the first two expressions of (5.76) with their counterparts in (5.67) we realize that, except for some minor differences in notation, they are identical. An important detail is that, since F does not contain u, the first two expressions are independent from the third. In other words, the characteristic system can be integrated first by solving the $2n$ equations involving just x_i and p_i, and only later solving the evolution equation for u. Notice also that, although in the PDE the symbols p_i stand for the partial derivatives $u_{,i}$, this fact is irrelevant as far as the characteristic equations are concerned.

In conclusion, a first-order PDE in reduced form gives rise, via its characteristic equations, to a Hamiltonian system!

Box 5.2 What is so special about a reduced first-order PDE?
We have called *reduced* a first-order PDE that does not explicitly include the unknown function u. According to Eq. (5.76) the characteristic equations of a reduced first-order PDE yield a Hamiltonian system, a remarkable fact considering that no Physics has been invoked. Remarkable too is the fact that every first order PDE can be brought into a reduced form, at a relatively small price. To see how this is possible, let us start from the general non-reduced PDE

$$F(x_1, \ldots, x_n, u, p_1, \ldots, p_n) = 0.$$

Instead of looking for a solution in the form $u = u(x_1, \ldots, x_n)$ let us look for a solution in the implicit form $w(x_1, \ldots x_n, u) = 0$, and let us investigate what PDE does the function w of $n + 1$ independent variables satisfy. Since

$$\frac{\partial w}{\partial x_1} dx_1 + \cdots + \frac{\partial w}{\partial x_n} dx_n + \frac{\partial w}{\partial u} du = 0,$$

we conclude that

$$p_i = \frac{\partial u}{\partial x_i} = -\frac{\frac{\partial w}{\partial x_i}}{\frac{\partial w}{\partial u}}.$$

Thus, the function w of the *independent* variables $x_1, \ldots, x_n, u$ satisfies the equation

$$F\left(x_1, \ldots, x_n, u, -\frac{\frac{\partial w}{\partial x_1}}{\frac{\partial w}{\partial u}}, \ldots, -\frac{\frac{\partial w}{\partial x_n}}{\frac{\partial w}{\partial u}}\right) = G\left(x_1, \ldots, x_n, u, \frac{\partial w}{\partial x_1}, \ldots, \frac{\partial w}{\partial x_n}, \frac{\partial w}{\partial u}\right) = 0$$

This is a first-order PDE in reduced form for the function w of $n + 1$ variables. Having obtained a solution of this equation, we have also obtained the solution u to the original equation in implicit form.

5.8.3 *The Hamilton–Jacobi Equation*

We are now in a position to answer the question as to whether, given a Hamiltonian system, one can always find a reduced first-order PDE whose characteristics are Hamilton's equations of the system. The answer is surprisingly simple, positive and constructive. Let $H = H(q_1, \ldots, q_n, p_1, \ldots, p_n, t)$ be the Hamiltonian of the system. The corresponding first-order PDE, known as the *Hamilton–Jacobi equation*, is

$$H\left(q_1, \ldots, q_n, \frac{\partial S}{\partial q_1}, \ldots, \frac{\partial S}{\partial q_n}, t\right) + \frac{\partial S}{\partial t} = 0. \tag{5.77}$$

The function $S = S(q_1, \ldots, q_n, t)$, obtained as a solution of this equation, is called an *action function*. The action function acts as some kind of scalar potential for the momenta.

5.8.4 An Example

The Hamilton–Jacobi equation establishes that any Hamiltonian system is ultimately governed by the solutions of a single scalar PDE of the first order. Apart from the theoretical importance of this result, it also has practical applications. Suppose we are given a Hamiltonian function H. The obvious way to analyze the evolution of the system out of some initial conditions is to solve the system of ODEs provided by Hamilton's equations. If, instead, we construct the associated Hamilton–Jacobi equation and if, by some technique, we manage to find a solution of this equation involving some arbitrary constants, then we have solved the problem in a completely different way. In particular, we can recover the characteristics corresponding to our initial conditions.

A *complete integral* of a first order PDE $F(x_1, \ldots, x_n, u, p_1, \ldots, p_n) = 0$ is a function $f(x_1, \ldots, x_n, a_1, \ldots, a_n)$ which satisfies the PDE for all values of the arbitrary parameters $a_1, \ldots, a_n$. The characteristics of the PDE are obtained from a complete integral by equating the derivatives of the complete integral with respect to each of the parameters to a constant.[8]

Box 5.3 Complete, general and singular integrals
We have learned how to solve the Cauchy problem for any particular initial data by the method of characteristics. Can we define, in some sense, a general solution? One way to approach this question is to note that, for the case of two independent variables x, y, the solution depends on the values of a function on a parametrized curve lying on the x, y plane. In other words, the solution depends on a function of a single variable. Keeping this observation in mind, let us define a *complete integral* of the PDE $F(x, y, u, p, q) = 0$ as a function

$$u = f(x, y, \alpha, \beta)$$

[8]See Boxes 5.3 and 5.4. For a thorough understanding of these topics within the mathematical context, see [1], p. 59, [2], p. 33, and [3], p. 29. For many interesting and challenging problems on the general integral, [4] is highly recommended.

that satisfies the PDE for *arbitrary* values of the two parameters α, β. Since the parameters are arbitrary and independent, we may decide to impose a restriction to the two-parameter family by choosing a specific functional dependence $\beta = \beta(\alpha)$. We have at our disposal an arbitrary function of a single variable to control the solution. More specifically, referring to Box 5.1, we can obtain the envelope of the new one-parameter family, and eliminate the parameter α, by choosing a specific function β and solving the algebraic system

$$u = f(x, y, \alpha, \beta(\alpha)) \qquad f_\alpha(x, y, \alpha, \beta(\alpha)) + f_\beta(x, y, \alpha, \beta(\alpha))\frac{d\beta(\alpha)}{d\alpha} = 0.$$

But, since, at each point, the envelope of a family coincides with one of the solutions and has the same tangent plane, we conclude that this envelop is itself a solution! We have thus obtained a solution depending on an arbitrary function β, that is, a *general integral*.

Finally, a *singular integral* can sometimes be found that is not comprised within the solutions delivered by the general integral. This singular solution is obtained as the envelope of the whole two-parameter family $u = f(x, y, \alpha, \beta)$. It can be regarded as an envelope of envelopes. It is delivered by the system of algebraic equations

$$u = f(x, y, \alpha, \beta) \qquad f_\alpha(x, y, \alpha, \beta) = 0 \qquad f_\beta(x, y, \alpha, \beta) = 0.$$

In order to read off α and β from the last two equations, according to the inverse function theorem, the Jacobian determinant $\partial(f_\alpha, f_\beta)/\partial(\alpha, \beta)$ (which happens to be the Hessian determinant) must not vanish.

For the sake of simplicity, we have only dealt with the case of two independent variables, but a similar treatment can be justified for higher dimensions.

Since complete integrals are in general difficult to obtain, let us deal with a simple example from Hamiltonian Mechanics, namely, the classical ballistic problem: A particle of mass m moving under the action of constant gravity g in the x, y plane, where y is the upright vertical direction. The Hamiltonian function in this case is the total energy, expressed in terms of coordinates x, y and momenta p, q as

$$H(x, y, p, q) = \frac{1}{2m}\left(p^2 + q^2\right) + mgy. \tag{5.78}$$

The Hamilton–Jacobi equation is, therefore,

$$\frac{1}{2m}\left(\left(\frac{\partial S}{\partial x}\right)^2 + \left(\frac{\partial S}{\partial y}\right)^2\right) + mgy + \frac{\partial S}{\partial t} = 0. \tag{5.79}$$

To find a complete integral we try a solution of the form

$$S(x, y, t) = f_1(x) + f_2(y) + f_3(t). \tag{5.80}$$

Substituting this assumption in Eq. (5.79) we obtain

$$\frac{1}{2m}\left(\frac{df_1}{dx}\right)^2 + \frac{1}{2m}\left(\frac{df_2}{dy}\right)^2 + mgy + \frac{df_3}{dt} = 0. \tag{5.81}$$

Since the only way that functions of different variables may be equated to each other is if they are constant, we obtain the three conditions

$$\frac{1}{2m}\left(\frac{df_1}{dx}\right)^2 = A \quad \frac{1}{2m}\left(\frac{df_2}{dy}\right)^2 + mgy = B \quad \frac{df_3}{dt} = -(A + B), \tag{5.82}$$

where A, B are arbitrary constants. Upon integration, we obtain

$$f_1(x) = \sqrt{2mA}\, x + k_1 \qquad f_2(y) = \frac{2\sqrt{2}(B - mgy)^{3/2}}{3gm^{1/2}} + k_2 \qquad f_3(t) = -(A + B)t + k_3, \tag{5.83}$$

where k_1, k_2, k_3 are constants of integration. The complete integral found is, therefore,

$$S(x, y, t, A, B, C) = \sqrt{2mA}\, x + \frac{2\sqrt{2}(B - mgy)^{3/2}}{3gm^{1/2}} - (A + B)t + C. \tag{5.84}$$

The constant C is irrelevant, since the function S can be determined only up to an additive constant as a result of the fact that the Hamilton–Jacobi equation is in reduced form (that is, S itself does not appear explicitly in the PDE). The characteristics are obtained by taking the derivatives of the complete integral with respect to the parameters and equating to constants, as explained in Box 5.4. We obtain

$$-t + \sqrt{\frac{m}{2A}}\, x = a \qquad -t + \sqrt{\frac{2(B - mgy)}{g^2 m}} = b. \tag{5.85}$$

As expected in the ballistic problem, the horizontal coordinate is a linear function of time, while the vertical coordinate varies quadratically. The four constants A, B, a, b can be pinned down when the initial conditions of position and momentum are specified.

Box 5.4 Obtaining the characteristics from the complete integral
We restrict attention to the case of a first-order PDE in reduced form with two independent variables, namely, $F(x, y, p, q) = 0$. Due to the assumed reduced form of the PDE, if $u = u(x, y)$ is a solution, so is $u = u(x, y) + \beta$, where β is an arbitrary constant. The immediate consequence of this fact is that, to form a complete integral, we need to find only a solution of the form $u = f(x, y, \alpha)$. The complete integral is given by $u = f(x, y, \alpha) + \beta$. Our intention is to obtain a general characteristic strip that can be fitted to any initial conditions. Consider the condition

$$f_\alpha(x, y, a) = \gamma,$$

where γ is a constant. We claim that, for any values of the three parameters α, β, γ, the four equations

$$f_\alpha(x, y, a) = \gamma \qquad u = f(x, y, \alpha) + \beta \qquad p = f_x(x, y, \alpha) \qquad q = f_y(x, y, \alpha)$$

define a characteristic strip of the PDE. We start by noticing that each of these equations eliminates, so to speak, a degree of freedom in the 5-dimensional space of coordinates x, y, u, p, q, so that 4 independent equations, in general, determine a curve in this space. We notice, moreover, that this line will lie on the 4-dimensional sub-manifold $F = 0$. This conclusion follows from the fact that the function $f(x, y, \alpha)$ is a solution of the PDE for arbitrary values of α. Thirdly, we verify that we have a three-parameter family of such (one-dimensional) curves sweeping the sub-manifold $F = 0$. Accordingly, to pin down one of these curves we need to adjust the parameters α, β, γ to satisfy any given initial conditions x_0, y_0, u_0, p_0, q_0. Clearly, q_0 is not independent of p_0, since the condition $F = 0$ must be satisfied by the initial conditions too. Finally, we verify that, for any fixed values of α, β, γ, the curve satisfies the strip condition (5.21). This is an immediate consequence of the fact that $du = f_x dx + f_y dy = pdx + qdy$.

All that is left is to ascertain that these strips are characteristic. Following [2], consider the differential of our main condition $f_\alpha(x, y, a) = \gamma$, that is,

$$f_{\alpha x} dx + f_{\alpha y} dy = 0.$$

This equation provides a specific ratio between dx and dy provided that the two partial derivatives do not vanish simultaneously. Substituting our complete integral into the PDE and differentiating with respect to α we obtain

$$F_p f_{x\alpha} + F_q f_{y\alpha} = 0.$$

Since mixed partial derivatives are symmetric, we obtain

$$\frac{dy}{dx} = \frac{F_q}{F_p}.$$

This result is equivalent to the first two conditions for a characteristic strip, as given in Eq. (5.23). The condition $du = pdx + qdy$, which we have already derived, implies that the third condition in (5.23) is fulfilled. The final two conditions are obtained comparing the differentials of p and q with the (vanishing) total derivatives of F with respect to x and y, respectively. For n independent variables, the parameter α is replaced by $n-1$ parameters α_i and the characteristics are obtained by equating the derivatives of the complete integral with respect to each of these parameters to a constant.

Exercises

Exercises 5.1 ([4], p. 66.) Find the characteristics of the equation $u_x u_y = u$ and determine the integral surface that passes through the parabola $x = 0,\ y^2 = u$.

Exercises 5.2 Modify the code of Box 3.3 to handle a general nonlinear first-order PDE in two independent variables.

Exercises 5.3 Solve the PDE of Example 5.1 in parametric form with the initial condition $u(x, y, 0) = 1/(1 + (x + y)^2)$. Plot the various profiles (as surfaces in $\mathbb{R}^3$) of the solution for several instants of time. Notice the formation of multiple-valued profiles, indicating the emergence of shock waves.

Exercises 5.4 Show that

$$\frac{\partial(x_1, \ldots, x_m)}{\partial(s, r_1, \ldots, r_{m-1})} = J,$$

where J is given by Eq. (5.72). [Hint: use Eq. (5.67)].

Exercises 5.5 Show explicitly how the characteristics of the Hamilton–Jacobi Eq. (5.77) reproduce Hamilton's equations (5.74). What happens when the Hamiltonian is independent of time?

References

1. Duff GFD (1956) Partial differential equations. Toronto University Press, Toronto
2. Garabedian PR (1964) Partial differential equations. Wiley, London
3. John F (1982) Partial differential equations. Springer, Berlin
4. Sneddon IN (1957) Elements of partial differential equations. McGraw-Hill, New York (Republished by Dover (2006))

Part III
Classification of Equations and Systems

Chapter 6
The Second-Order Quasi-linear Equation

A careful analysis of the single quasi-linear second-order equation is the gateway into the world of higher-order partial differential equations and systems. One of the most important aspects of this analysis is the distinction between hyperbolic, parabolic and elliptic types. From the physical standpoint, the hyperbolic type corresponds to physical systems that can transmit sharp signals over finite distances. The parabolic type represents diffusive phenomena. The elliptic type is often associated with statical situations, where time is absent. Of these three types, the hyperbolic case turns out to resemble the single first-order PDE the most. In particular, characteristic lines make their appearance and play a role in the understanding of the propagation phenomena and in the prediction of the speed, trajectory and variation in amplitude of the signals, without having to solve the differential equation itself.

6.1 Introduction

The general form of a quasi-linear second-order PDE for a function $u = u(x, y)$ of two independent variables is

$$a\, u_{xx} + 2b\, u_{xy} + c\, u_{yy} = d\ , \tag{6.1}$$

where a, b, c, d are functions of the arguments x, y, u, u_x, u_y. Recall that by quasi-linearity we mean that we only require that the highest derivatives in the equation appear linearly. Both in the case of ODEs and in the case of first-order PDEs, we had a chance of appreciating the importance of interpreting a differential equation, at least locally, in the following way: If we are given appropriate initial conditions on an appropriate set (a point, a line), the differential equation gives us information on how to *come out*, as it were, from this initial set. The nature of the initial set and the nature of the initial information that needs to be given on it depend on the type of problem

M. Epstein, *Partial Differential Equations*, Mathematical Engineering,
DOI 10.1007/978-3-319-55212-5_6

and on the type of equation. For example, in the case of the dynamics of a system of n particles moving in space, the equations of motion are $3n$ second-order (or, equivalently, $6n$ first-order) ordinary differential equations. If, at some initial time, we prescribe the instantaneous positions and velocities of each particle, a total of $6n$ numbers, the differential equations allow us to calculate the accelerations, and thus to come out of the original state and predict the evolution of the system. Notice that, by successive differentiations of the equations of motion, if all the functions involved are analytic (i.e., if they admit a convergent Taylor expansion), we can obtain any number of derivatives at the initial time. In this way, at least for the analytic case, we can extend the solution to a finite interval of time.[1]

In the case of a single first-order PDE we have seen that the Cauchy problem consists of specifying the values of the unknown function on an initial curve. In essence, unless the curve happens to be a characteristic of the PDE, the differential equation allows us to come out of this curve using the information provided by the differential equation concerning the first derivatives. Again we observe that the information is given on a set with one dimension less than the space of independent variables and it involves knowledge of the function and its derivatives up to and including one degree less than the order of the PDE. In the case of a first-order equation, one degree less means no derivatives at all. It is natural, therefore, to expect that the Cauchy problem for a second-order PDE in two dimensions will involve specifying, on a given curve, the unknown function and its first derivatives. We expect then the second-order PDE to provide enough information about the second (and higher) derivatives so that we can come out of the initial curve. In fact, the classical theorem of Cauchy–Kowalewski[2] proves that if the coefficients in the PDE are analytic, this procedure can be formalized to demonstrate the existence and uniqueness of the (analytic) solution in some neighbourhood of a point on the initial manifold. That said, we don't want to convey the impression that this *initial-value problem* is prevalent in the treatment and application of all possible equations. We will see later that boundary value problems and mixed initial-boundary-value problems are prevalent in applications. But, from the conceptual point of view, the understanding of the behaviour of the equation and its solution in the vicinity of an initial manifold with known initial data is of paramount importance. In particular, we will presently see how it can be used to classify the possible second-order quasi-linear equations into three definite types.[3]

---

[1] The theorem of existence and uniqueness, on the other hand, requires much less than analyticity.

[2] For a detailed proof, see the classical treatise [1].

[3] In some textbooks, the classification is based on the so-called normal forms of the equations. In order to appreciate the meaning of these forms, however, it is necessary to have already seen an example of each. We prefer to classify the equations in terms of their different behaviour vis-à-vis the Cauchy problem. Our treatment is based on [4], whose clarity and conciseness are difficult to match.

6.2 The First-Order PDE Revisited

Before proceeding to the study of the second-order equation, it may be useful to revisit the first-order case. Given the first-order quasi-linear PDE

$$a(x, y, u)\, u_x + b(x, y, u)\, u_y = c(x, y, u), \tag{6.2}$$

and a parametrized (initial) curve

$$x = \hat{x}(r) \qquad y = \hat{y}(r), \tag{6.3}$$

in the space of independent variables on which the value of the solution is specified as

$$u = \hat{u}(r), \tag{6.4}$$

the Cauchy problem consists of finding an integral surface that contains the space curve represented by Eqs. (6.3) and (6.4). Let us assume that we know nothing at all about characteristics and their role in building integral surfaces. We ask the question: Does the differential equation (6.2) provide us with enough information about the first derivatives of the solution being sought so that we can come out, as it were, of the initial curve? Intuitively, the answer will be positive if, and only if, the PDE contains information about the directional derivative in a direction transversal to the original curve in the plane. We may ask why, if we need the whole gradient, we only demand the derivative in one transversal direction. The answer is: because the initial data, as given by Eq. (6.4), already give us the necessary information in the direction of the curve itself. Indeed, assuming that the function $\hat{u}$ is differentiable, we obtain by the chain rule of differentiation

$$\frac{d\hat{u}}{dr} = u_x\, \frac{d\hat{x}}{dr} + u_y\, \frac{d\hat{y}}{dr}, \tag{6.5}$$

where u_x and u_y are the partial derivatives of any putative solution of the PDE. This equation must hold true along the whole initial curve, if indeed we want our solution to satisfy the given initial conditions. Always moving along the initial curve, we see that the determination of the two partial derivatives u_x, u_y at any given point along this curve is a purely algebraic problem, consisting in solving (point-wise) the linear system of equations

$$\begin{bmatrix} a & b \\ \hat{x}' & \hat{y}' \end{bmatrix} \begin{Bmatrix} u_x \\ u_y \end{Bmatrix} = \begin{Bmatrix} c \\ \hat{u}' \end{Bmatrix}, \tag{6.6}$$

where we indicate by a prime the derivative with respect to the curve parameter. The first equation of this linear system is provided by the PDE and the second equation of the system is provided by the initial conditions via Eq. (6.5). No more information is available. This linear system will have a unique solution if, and only if, the determinant of the coefficient matrix does not vanish. If this is the case, we obtain in a unique fashion the whole gradient of the solution and we can proceed to extend the initial data to a solution in the nearby region (for example, by a method of finite differences). Otherwise, namely if the determinant vanishes, there are two possibilities:

1. The rank of the augmented matrix

$$\begin{bmatrix} a & b & c \\ \hat{x}' & \hat{y}' & \hat{u}' \end{bmatrix} \tag{6.7}$$

   is 2, in which case there is no solution;
2. The rank of this augmented matrix is less than 2, in which case there is an infinite number of solutions.

In this way, we discover the concept of characteristic direction in a natural way as an answer to the question: When does the Cauchy problem not have a unique solution in the neighbourhood of a point? The answer can be interpreted as follows: When the PDE itself does not add any information to that provided by the initial data.

6.3 The Second-Order Case

Inspired by the analysis of the first-order case, we formulate the Cauchy problem for the second-order quasi-linear PDE (6.1) as follows: Given a curve by Eq. (6.3), and given, along this curve, the (initial) values of the unknown function and its first partial derivatives

$$u = \hat{u}(r), \tag{6.8}$$

$$u_x = \hat{u}_1(r), \tag{6.9}$$

$$u_y = \hat{u}_2(r), \tag{6.10}$$

where $\hat{u}(r)$, $\hat{u}_1(r)$ and $\hat{u}_2(r)$ are differentiable functions of the single variable r, find a solution of Eq. (6.1) compatible with these Cauchy data.

Before investigating whether or not this problem has a solution, it is worth remarking that the functions $\hat{u}_1(r)$ and $\hat{u}_2(r)$ cannot be specified arbitrarily. Indeed, they must be compatible with the derivative of the function $\hat{u}(r)$, as we have seen above in Eq. (6.5). Specifically,

$$\hat{u}_1 \frac{d\hat{x}}{dr} + \hat{u}_2 \frac{d\hat{y}}{dr} = \frac{d\hat{u}}{dr}. \tag{6.11}$$

An equivalent way to prescribe the data needed for the Cauchy problem, and to avoid a contradiction, is to stipulate the value of the unknown function u on the initial curve and the value of the first derivative du/dn in any direction $\mathbf{n}$ *transversal* to the curve.

By analogy with the first-order case, we want to investigate whether or not the differential equation provides us with enough information about the second derivatives so that we can come out of the initial curve with the given initial data. For this to be possible, we need to be able to calculate at each point of the initial curve the values of all three second partial derivatives. We start by remarking that the Cauchy data already provide us with some information about the second derivatives of any proposed solution, just like in the first-order case. Indeed, by the chain rule of differentiation, we know that at any point of the initial curve

$$\hat{u}_1' = u_{xx}\,\hat{x}' + u_{xy}\,\hat{y}' \tag{6.12}$$

and

$$\hat{u}_2' = u_{yx}\,\hat{x}' + u_{yy}\,\hat{y}'. \tag{6.13}$$

Thus, the determination of the three second partial derivatives of the solution along the initial curve is, at each point, a purely algebraic problem defined by the system of linear equations

$$\begin{bmatrix} a & 2b & c \\ \hat{x}' & \hat{y}' & 0 \\ 0 & \hat{x}' & \hat{y}' \end{bmatrix} \begin{Bmatrix} u_{xx} \\ u_{xy} \\ u_{yy} \end{Bmatrix} = \begin{Bmatrix} d \\ \hat{u}_1' \\ \hat{u}_2' \end{Bmatrix} \tag{6.14}$$

The determinant of the coefficient matrix of this system is given by

$$\Delta = a\,\hat{y}'^2 - 2b\,\hat{x}'\,\hat{y}' + c\,\hat{x}'^2. \tag{6.15}$$

If this determinant does not vanish at any point along the initial curve (for the given Cauchy data), there exists a point-wise unique solution for all three second partial derivatives and, therefore, it is possible to come out of the initial curve by means of the information gathered from the initial data and the differential equation. Otherwise, that is, when the determinant vanishes, if the rank of the augmented matrix is equal to 3, the system is incompatible and there are no solutions. If this rank is less than 3, the system is compatible, but the solution is not unique. In this case, if the determinant vanishes identically along the initial curve, the curve is called a characteristic of the differential equation for the given Cauchy data. Notice that in the second-order case we will be representing only the projected curves on the x, y plane, since the visualization of the entire Cauchy problem as a curve would involve a space of 5 dimensions. If the problem happens to be linear, the projected characteristics are independent of the initial data.

According to what we have just learned, in the linear case the characteristic curves can be seen as solutions of the ODE

$$a\,\hat{y}'^2 - 2b\,\hat{x}'\,\hat{y}' + c\,\hat{x}'^2 = 0, \tag{6.16}$$

which (assuming, for example, that $\hat{x}' \neq 0$) can be written as

$$a\left(\frac{dy}{dx}\right)^2 - 2b\left(\frac{dy}{dx}\right) + c = 0. \tag{6.17}$$

If, on the other hand, the PDE is quasi-linear, the coefficients of this equation may be functions of u and its two (first) partial derivatives. What this means is that, in the legitimate quasi-linear case, we are dealing with characteristics that depend on the initial data, just as in the first-order case. At any rate, Eq. (6.17) reduces to

$$\frac{dy}{dx} = \frac{b \pm \sqrt{b^2 - ac}}{a}. \tag{6.18}$$

We have assumed that $a \neq 0$, for the sake of the argument. We see that in the case of second order equations, in contradistinction with the first-order case, there may be no characteristics at all! This occurs when the discriminant of the quadratic equation happens to be negative, namely when

$$b^2 - a\,c < 0. \tag{6.19}$$

If this is the case, the equation is called *elliptic* at the point in question and for the given initial data. At the other extreme, we have the case in which two distinct characteristics exist. This happens when the discriminant is positive, that is,

$$b^2 - a\,c > 0. \tag{6.20}$$

In this case, the equation is called *hyperbolic* at the point in question and for the given initial data. The intermediate case, when

$$b^2 - a\,c = 0, \tag{6.21}$$

is called *parabolic*. In this case, we have just one characteristic direction.[4] The reason for this terminology, as you may have guessed, is that quadratic forms in two variables give rise to ellipses, hyperbolas or parabolas precisely according to the above criteria. If the original PDE is not only linear but also with constant coefficients, then the type (elliptic, hyperbolic or parabolic) is independent of position and of the solution. If the equation is linear, but with variable coefficients, the type is independent of the solution, but it may still vary from point to point. For the truly quasi-linear case, the type may depend both on the position and on the solution. In light of the second-order case, we can perhaps say that the single first-order PDE is automatically hyperbolic.

[4]Although we have assumed that the first coefficient of the PDE does not vanish, in fact the conclusion that the number of characteristic directions is governed by the discriminant of the quadratic equation is valid for any values of the coefficients, provided, of course, that not all three vanish simultaneously.

Although this is not a precise statement, it is indeed the case that the treatment of hyperbolic second-order PDEs is, among all three types, the one that most resembles the first-order counterpart in terms of such important physical notions as the ability to propagate discontinuities.

6.4 Propagation of Weak Singularities

6.4.1 Hadamard's Lemma and Its Consequences

Hadamard's lemma[5] is a theorem of calculus that legitimizes the calculation of the directional derivative of a function by means of the chain rule when the desired direction lies on the boundary, rather than just in the interior, of the domain of definition. To be more precise, let $\mathcal{D}$ be an open domain in $\mathbb{R}^n$ and let ψ be a function

$$\psi : \mathcal{D} \to \mathbb{R} \tag{6.22}$$

of class C^1. Put differently, ψ is a real valued function of n independent real variables $x_1, \ldots, x_n$ with continuous first derivatives $\psi_{,i}$ $(i = 1, \ldots, n)$. Let us, moreover, assume that the boundary $\partial\mathcal{D}$ of $\mathcal{D}$ is smooth and that, as we approach the boundary along any interior path, the function ψ and each of its partial derivatives $\psi_{,i}$ $(i = 1, \ldots, n)$ approach finite limits $\bar{\psi}$ and $\bar{\psi}_{,i}$, respectively. Consider a smooth curve

$$x_i = x_i(s) \qquad (i = 1, \ldots, n) \tag{6.23}$$

lying on the boundary $\partial\mathcal{D}$. Then, Hadamard's lemma asserts that

$$\frac{d\bar{\psi}}{ds} = \sum_{i=1}^{n} \bar{\psi}_{,i} \frac{dx_i}{ds}. \tag{6.24}$$

This result may not look spectacular, since it is what we would have done anyway, without asking for Hadamard's permission, but it does allow us to use the chain rule even when the function is not defined over an open (tubular) neighbourhood containing the curve. Why is this important at this point of our treatment? The reason is as follows. Let us assume that we have a domain $\mathcal{D} \subset \mathbb{R}^n$ subdivided into two sub-domains, $\mathcal{D}^+$ and $\mathcal{D}^-$, as shown in Fig. 6.1, whose boundaries share a common smooth part Λ, a manifold of dimension $n - 1$, also called a *hyper-surface* or just a *surface*.

[5]The treatment in this section draws from [5], pp. 491–529. It should be pointed out that Hadamard proved several theorems and lemmas that carry his name. The lemma used in this section is a rather elementary result in Calculus. Its proof was originally given by Hadamard in [3], p. 84.

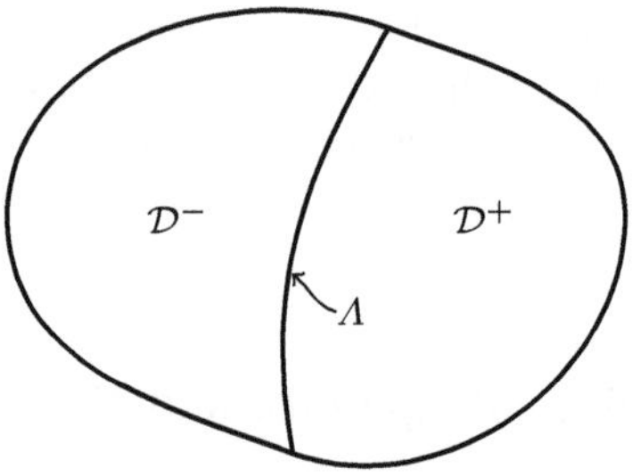

Fig. 6.1 A surface of discontinuity

Assume that the function ψ and each of its derivatives $\psi_{,i}$ are continuous in the interior of each of the respective sub-domains, but that they attain possibly different limits, $(\psi^-, \psi^-_{,i})$ and $(\psi^+, \psi^+_{,i})$, as we approach Λ, according to whether we come from paths within $\mathcal{D}^-$ or $\mathcal{D}^+$, respectively. In other words, the given function and/or its derivatives undergo a jump upon crossing Λ. We refer to Λ as a *singular surface* or a *surface of discontinuity*. We will use the following convenient short-hand notation to denote the jump of a quantity such as ψ across Λ:

$$[\![\psi]\!] = \psi^+ - \psi^-. \tag{6.25}$$

According to Hadamard's lemma, we are in a position to apply the chain rule (6.24) independently at each of the sub-domains to calculate the derivative of ψ along a smooth curve lying on Λ, namely,

$$\frac{d\psi^+}{ds} = \sum_{i=1}^{n} \psi^+_{,i} \frac{dx_i}{ds} \quad \text{and} \quad \frac{d\psi^-}{ds} = \sum_{i=1}^{n} \psi^-_{,i} \frac{dx_i}{ds}. \tag{6.26}$$

Subtracting the second equation from the first and using the notation (6.25) we obtain

$$\frac{d[\![\psi]\!]}{ds} = \sum_{i=1}^{n} [\![\psi_{,i}]\!] \frac{dx_i}{ds}. \tag{6.27}$$

In other words, the derivative of the jump of a function in a direction tangential to the surface of discontinuity is given by the jump of the derivative in the same direction. Thus, the jumps of a function and of its partial derivatives *cannot be entirely arbitrary*, but must be related by Eq. (6.27). It is interesting to note that in the case for which the function ψ happens to be continuous across Λ, the jumps of its derivatives must satisfy the condition

$$\sum_{i=1}^{n} [\![\psi_{,i}]\!] \frac{dx_i}{ds} = 0 \tag{6.28}$$

This condition can be stated as follows: If ψ is continuous across Λ, the jump of its gradient is orthogonal to Λ.[6]

Returning to the general case, if we were to choose a local coordinate system with all but one of its natural base vectors lying on Λ, the derivative in the direction of the transverse coordinate would not be at all involved in condition (6.27), as one would intuitively expect. Equation (6.27) embodies the so-called *geometric compatibility conditions*. This terminology is meant to emphasize the fact that these conditions emerge from a purely geometric analysis of the situation, without any reference to equations of balance that may arise from the physical formulation of a particular problem.

When one of the independent variables of the problem is identified with time and the remaining ones with space, a singular surface can be regarded as the propagation of a *wave front*. In the case of just two independent variables, the wave front consists of a single point. The slope of the singular curve can, in this case, be seen as the speed of propagation of the front. This idea can be generalized for the case of more than two independent variables.

6.4.2 Weak Singularities

Given a PDE of order n, a singular surface is said to be *weak* if only the n-th (or higher) derivatives are discontinuous across it, while the lower derivatives are all continuous. For a second-order equation, for example, a weak singular surface may only carry discontinuities of the derivatives of order 2 and higher. In Continuum Mechanics applications, where the relevant equations are indeed of second order, these singularities are known as *acceleration waves*. If the first derivatives are discontinuous, we are in the presence of *strong singularities*, or *shocks*. If the function itself (the displacement, say) is discontinuous, we are in the presence of a dislocation. This terminology may not apply in other contexts.

Consider the general quasi-linear second-order PDE (6.1). Taking the jump of this equation across a weak singular curve with parametric equations

$$x = \tilde{x}(s) \qquad y = \tilde{y}(s), \tag{6.29}$$

we obtain along this curve the jump condition

$$a\, [\![u_{xx}]\!] + 2b\, [\![u_{xy}]\!] + c\, [\![u_{yy}]\!] = 0. \tag{6.30}$$

[6] In using this terminology, we are implicitly assuming that we have defined the natural dot product in $\mathbb{R}^n$. A more delicate treatment, would consider the gradient not as a vector but as a differential form which would then be annihilated by vectors forming a basis on the singular surface. We have already discussed a similar situation in Box 3.2.

We have assumed that the coefficients of the PDE (for example, the moduli of elasticity) are smooth functions throughout the domain of interest. Since we are dealing with a weak singularity, all the first derivatives are continuous across the singular curve, implying that Eq. (6.28) may be applied, identifying the generic function ψ successively with each of the two first-derivatives. As a result, we obtain the two conditions

$$[\![u_{xx}]\!]\,\frac{d\tilde{x}}{ds} + [\![u_{xy}]\!]\,\frac{d\tilde{y}}{ds} = 0, \tag{6.31}$$

$$[\![u_{yx}]\!]\,\frac{d\tilde{x}}{ds} + [\![u_{yy}]\!]\,\frac{d\tilde{y}}{ds} = 0. \tag{6.32}$$

Equations (6.30), (6.31) and (6.32) can be written in matrix form as

$$\begin{bmatrix} a & 2b & c \\ \tilde{x}' & \tilde{y}' & 0 \\ 0 & \tilde{x}' & \tilde{y}' \end{bmatrix} \begin{Bmatrix} [\![u_{xx}]\!] \\ [\![u_{xy}]\!] \\ [\![u_{yy}]\!] \end{Bmatrix} = \begin{Bmatrix} 0 \\ 0 \\ 0 \end{Bmatrix}, \tag{6.33}$$

where primes indicate derivatives with respect to the curve parameter. For this homogeneous system of linear equations to have a non-trivial solution, its determinant must vanish. In this way, we recover condition (6.16), implying that *weak discontinuities can only exist on characteristic curves*! If one of the independent variables is time, this conclusion can be expressed as the fact that weak signals propagate along characteristic curves, and their (local) speed of propagation is measured by the slope of these curves. Notice that, in particular, *elliptic equations cannot sustain weak discontinuities*, since they have no characteristic curves.

Note that Eqs. (6.31) and (6.32) imply that the jumps of the second derivatives are all interrelated. Denote, for example,

$$[\![u_{xx}]\!] = B. \tag{6.34}$$

Then, we obtain directly from (6.31) and (6.32)

$$[\![u_{xy}]\!] = -B\frac{d\tilde{x}}{d\tilde{y}} \tag{6.35}$$

and

$$[\![u_{yy}]\!] = B\left(\frac{d\tilde{x}}{d\tilde{y}}\right)^2. \tag{6.36}$$

In Continuum Mechanics applications this result implies that discontinuities in the acceleration are accompanied by discontinuities in the strain rate of the deforming medium. A more general version of Eqs. (6.34)–(6.36), which will be used later, is presented in Box 6.1.

Box 6.1 Iterated compatibility conditions

Suppose that a function $\psi = \psi(x_1, \ldots, x_n)$ and all its first partial derivatives are continuous across Λ. By Eq. (6.28), at any given point on Λ

$$[\![\psi_{,ij}]\!] = [\![(\psi_{,i})_{,j}]\!] = a_i n_j,$$

where n_i are the (Cartesian) components of a vector perpendicular to the hypersurface Λ and a_i is a coefficient of proportionality. These quantities constitute a vector $\mathbf{a}$. But we can also write

$$[\![\psi_{,ji}]\!] = [\![(\psi_{,j})_{,i}]\!] = a_j n_i.$$

By the equality of mixed partial derivatives we conclude that

$$a_i n_j = a_j n_i.$$

This equality is only possible if the vectors $\mathbf{a}$ and $\mathbf{n}$ are collinear. We conclude that there exists a scalar μ such that $a_i = \mu n_i$ for all $i = 1, \ldots, n$. We can, therefore, write

$$[\![\psi_{,ij}]\!] = \mu n_i n_j.$$

This elegant general result is known as the *iterated geometric compatibility condition*. It can be generalized for higher derivatives.

6.4.3 *Growth and Decay*

It is, in fact, possible to squeeze out more information. For simplicity, we will confine attention to the homogeneous linear case so that, in particular, the coefficients a, b, c are independent of the unknown function and its derivatives, while the right-hand side vanishes.[7] Assume that the singularity curve is nowhere tangential to the x direction. Differentiating the PDE with respect to x and then taking jumps we obtain

$$a\,[\![u_{xxx}]\!] + 2b\,[\![u_{xxy}]\!] + c\,[\![u_{xyy}]\!] = a_x\,[\![u_{xx}]\!] + 2b_x\,[\![u_{xy}]\!] + c_x\,[\![u_{yy}]\!] = 0. \quad (6.37)$$

The jumps of the third derivatives, however, are not independent of those of the second derivatives, as we know from the geometric compatibility conditions (6.27). Indeed,

[7] For the treatment of the quasi-linear case, see Box 6.2.

identifying the generic function ψ successively with each of the second derivatives, we can write

$$\frac{[\![u_{xx}]\!]}{ds} = [\![u_{xxx}]\!]\,\tilde{x}' + [\![u_{xxy}]\!]\,\tilde{y}', \tag{6.38}$$

$$\frac{[\![u_{xy}]\!]}{ds} = [\![u_{xyx}]\!]\,\tilde{x}' + [\![u_{xyy}]\!]\,\tilde{y}', \tag{6.39}$$

$$\frac{[\![u_{yy}]\!]}{ds} = [\![u_{yyx}]\!]\,\tilde{x}' + [\![u_{yyy}]\!]\,\tilde{y}'. \tag{6.40}$$

Multiplying Eqs. (6.38) and (6.39), respectively, by $a\,\tilde{y}'$ and $c\,\tilde{x}'$, and then adding the results we obtain

$$a\,\tilde{y}'\,\frac{[\![u_{xx}]\!]}{ds} + c\,\tilde{x}'\,\frac{[\![u_{xy}]\!]}{ds} = \left(a\,[\![u_{xxx}]\!] + c\,[\![u_{xyy}]\!]\right)\tilde{x}'\,\tilde{y}' + [\![u_{xxy}]\!]\left(a\,\tilde{y}'^2 + C\,\tilde{x}'^2\right). \tag{6.41}$$

Since the curve in question is characteristic, we can apply Eq. (6.16) to the last term of Eq. (6.41), thus yielding

$$a\,\tilde{y}'\,\frac{[\![u_{xx}]\!]}{ds} + c\,\tilde{x}'\,\frac{[\![u_{xy}]\!]}{ds} = \left(a\,[\![u_{xxx}]\!] + 2b\,[\![u_{xxy}]\!] + c\,[\![u_{xyy}]\!]\right)\tilde{x}'\,\tilde{y}'. \tag{6.42}$$

Introducing this result into Eq. (6.37), we obtain

$$a\,\tilde{y}'\,\frac{[\![u_{xx}]\!]}{ds} + c\,\tilde{x}'\,\frac{[\![u_{xy}]\!]}{ds} + \left(a_x\,[\![u_{xx}]\!] + 2b_x\,[\![u_{xy}]\!] + c_x\,[\![u_{yy}]\!]\right)\tilde{x}'\,\tilde{y}' = 0. \tag{6.43}$$

We have succeeded in obtaining an equation relating exclusively the jumps of the second derivatives and their derivatives with respect to the curve parameter. By virtue of Eqs. (6.34), (6.35) and (6.36), we can write

$$a\,\tilde{y}'\,\frac{dB}{ds} - c\,\tilde{x}'\,\frac{d(B\tilde{x}'/\tilde{y}')}{ds} + \left(a_x - 2b_x\,(\tilde{x}'/\tilde{y}') + c_x\,(\tilde{x}'/\tilde{y}')^2\right)B\,\tilde{x}'\,\tilde{y}' = 0. \tag{6.44}$$

This is a first-order ODE for the evolution of the magnitude of the jump. It is sometimes called the *transport equation* or the *decay-induction equation*.

If the characteristic curve is parametrized by y, which can always be done locally around a point at which $\tilde{y}' \neq 0$, the transport equation can be written more compactly as

$$a\,\frac{dB}{dy} - c\,\lambda\,\frac{d(B\,\lambda)}{dy} + \left(a_x - 2b_x\,\lambda + c_x\,\lambda^2\right)B = 0, \tag{6.45}$$

where λ denotes the characteristic slope $\tilde{x}'/\tilde{y}'$. Given that, except for $B = B(y)$, all the quantities involved are known from the solution of the characteristic equation, the integration of the transport equation is elementary.

Box 6.2 The transport equation for the quasi-linear case
If the PDE is quasi-linear, at least one of the coefficients a, b, c depends on at least one of the first derivatives of u. For the sake of the argument, suppose that the first term of the equation is of the form $a(x, y, u, u_x, u_y)u_{xx}$. As we reach the step in the derivation of the transport equation corresponding to Eq. (6.37), we need to calculate the derivative of this term with respect to x and then take the jump of the result, which leads to

$$\begin{aligned}[\![(a\, u_{xx})_x]\!] &= [\![\left(a_x + a_u u_x + a_{u_x} u_{xx} + a_{u_y} u_{yx}\right) u_{xx} + a u_{xxx}]\!] \\ &= (a_x + a_u u_x)[\![u_{xx}]\!] + a_{u_x}[\![u_{xx}^2]\!] + a_{u_y}[\![u_{yx} u_{xx}]\!] + a[\![u_{xxx}]\!].\end{aligned}$$

Thus we encounter the jump of products of two quantities, say p and q. A straightforward calculation delivers

$$[\![pq]\!] = -[\![p]\!][\![q]\!] + p^+[\![q]\!] + q^+[\![p]\!].$$

There are two immediate consequences of this formula. The first is that the transport equation becomes a non-linear ODE. The second consequence, bearing an important physical repercussion, is that, as the wave front advances and encounters preexisting values $u_{xx}^+, u_{xy}^+, u_{yy}^+$ ahead of the wave, these values affect the decay or growth of the amplitude of the propagating discontinuity.

6.5 Normal Forms

It is a useful exercise, that we could have carried out also for first-order equations, to ask ourselves how the form of a PDE is affected by an arbitrary change of coordinates. A new system of coordinates in the plane is specified by means of two smooth functions of two new variables ξ and η, namely,

$$x = x(\xi, \eta) \qquad y = y(\xi, \eta). \tag{6.46}$$

For this coordinate change to be legitimate within a certain region (a *coordinate patch*), we must also require that within this region the Jacobian determinant

$$J = \frac{\partial(x, y)}{\partial(\xi, \eta)} = \begin{vmatrix} x_\xi & x_\eta \\ y_\xi & y_\eta \end{vmatrix} \tag{6.47}$$

must not vanish at any point. Only when this is the case, which we assume from here on, can Eq. (6.46) be inverted to yield ξ, η as (smooth) functions of x, y. The function $u = u(x, y)$ can be expressed in terms of the new variables by composition of functions, namely,

$$u = u(x, y) = u(x(\xi, \eta), y(\xi, \eta)) = \hat{u}(\xi, \eta). \tag{6.48}$$

We are trying to distinguish, by means of a hat over the symbol, between the function as an operator and the result of applying this operator to the arguments. When there is no room for confusion, however, this practice can be abandoned and let the context indicate which is the function being considered. By a direct iterated use of the chain rule of differentiation, we obtain the expressions

$$\begin{aligned}
u_x &= \hat{u}_\xi \, \xi_x + \hat{u}_\eta \, \eta_x, \\
u_y &= \hat{u}_\xi \, \xi_y + \hat{u}_\eta \, \eta_y, \\
u_{xx} &= \hat{u}_{\xi\xi} \, \xi_x^2 + 2\hat{u}_{\xi\eta} \, \xi_x \, \eta_x + \hat{u}_{\eta\eta} \, \eta_x^2, \\
u_{xy} &= \hat{u}_{\xi\xi} \, \xi_x \, \xi_y + 2\hat{u}_{\xi\eta} \, (\xi_x \, \eta_y + \xi_y \, \eta_x) + \hat{u}_{\eta\eta} \, \eta_x \, \eta_y, \\
u_{yy} &= \hat{u}_{\xi\xi} \, \xi_y^2 + 2\hat{u}_{\xi\eta} \, \xi_y \, \eta_y + \hat{u}_{\eta\eta} \, \eta_y^2.
\end{aligned} \tag{6.49}$$

In the new coordinate system, therefore, the original PDE (6.1) can be written as

$$\hat{a} \, \hat{u}_{\xi\xi} + 2\hat{b} \, \hat{u}_{\xi\eta} + \hat{c} \, \hat{u}_{\eta\eta} = \hat{d}. \tag{6.50}$$

In this expression, we assume that the arguments of the coefficients have been expressed in terms of the new variables. The new coefficients of the second-order terms are given by the quadratic expressions

$$\begin{aligned}
\hat{a} &= a \, \xi_x^2 + 2b \, \xi_x \, \xi_y + c \, \xi_y^2, \\
\hat{b} &= a \, \xi_x \, \eta_x + b \, (\xi_x \, \eta_y + \xi_y \, \eta_x) + c \, \xi_y \, \eta_y, \\
\hat{c} &= a \, \eta_x^2 + 2b \, \eta_x \, \eta_y + c \, \eta_y^2.
\end{aligned} \tag{6.51}$$

The second-order part of the original PDE can be regarded as a quadratic form governed by the matrix

$$A = \begin{bmatrix} a & b \\ b & c \end{bmatrix}. \tag{6.52}$$

It is not difficult to verify that the counterpart for the transformed equation is governed by the matrix

$$\hat{A} = \begin{bmatrix} \hat{a} & \hat{b} \\ \hat{b} & \hat{c} \end{bmatrix} = J^{-1} \, A \, J^{-T}, \tag{6.53}$$

where J stands now for the Jacobian matrix. Notice that the determinants of both matrices, A and $\hat{A}$, will always have the same sign or vanish simultaneously. Each of these determinants is precisely the discriminant of the quadratic equation that we used to distinguish between the three types of equations. As expected, therefore, if a second-order quasi-linear PDE is hyperbolic (parabolic, elliptic) in one coordinate system, it will remain hyperbolic (parabolic, elliptic) in any other. Thus, the equation type is an invariant quality and carries a definite physical meaning. An argument based, among other considerations, on the algebra of symmetric real matrices, can be used to show[8] that, with a suitable change of coordinates, a quasi-linear hyperbolic equation can always be expressed, in some coordinate chart, as

$$\hat{u}_{\xi\eta} + \ldots = 0, \tag{6.54}$$

where we have indicated only the principal (i.e., second-order) part. An alternative coordinate transformation can be found that brings the hyperbolic equation to the form

$$\hat{u}_{\xi\xi} - \hat{u}_{\eta\eta} + \ldots = 0. \tag{6.55}$$

These are the so-called *normal forms* of an equation of the hyperbolic type.

For a parabolic equation, the normal form is

$$\hat{u}_{\xi\xi} + \ldots = 0, \tag{6.56}$$

Finally, for the elliptic equation we have the normal form

$$\hat{u}_{\xi\xi} + \hat{u}_{\eta\eta} + \ldots = 0, \tag{6.57}$$

The existence of these normal forms prompts us to study, in separate chapters, three paradigmatic equations, one of each type, corresponding to the simplest possible forms of the omitted terms. These are, respectively, the wave equation and the heat equation in one spatial dimension and the Laplace equation in two spatial dimensions.

Exercises

Exercise 6.1 Show that option 2 after Eq. (6.7), if applied to every point along the initial curve, corresponds exactly to the fact that the space curve represented by Eqs. (6.3) and (6.4) is a characteristic curve of the PDE (6.2).

Exercise 6.2 Show that the first-order quasi-linear PDE (6.2) can be regarded as the specification of the directional derivative of the unknown function in the characteristic direction.

Exercise 6.3 For each of the following second-order PDEs determine the type (elliptic, parabolic, hyperbolic) and, if necessary, the regions over which each type applies. Obtain and draw the characteristic curves wherever they exist.

---

[8]The proof is not as straightforward as it may appear from the casual reading of some texts. A good treatment of these normal or *canonical* forms can be found in [2].

$$2.5u_{xx} + 5u_{xy} + 1.5u_{yy} + 5u = 0.$$

$$2u_{xx} + 4u_{xy} + 2u_{yy} + 3(x^2 + y^2)u_x = e^y.$$

$$u_{xx} - 2u_{xy} + 2u_{yy} + 4u_x u_y = 0.$$

$$u_{xx} + x\ u_{yy} = 0. \qquad \text{(Tricomi equation)}$$

Exercise 6.4 For a solid body without cracks in the small-deformation regime, find which components of the strain tensor may be discontinuous across some plane. [Hint: the displacement field is continuous].

Exercise 6.5 Generalize the transport equation (6.44) to the case of the non-homogeneous linear equation (6.1), where

$$d = e(x, y)\ u_x + f(x, y)\ u_y + g(x, y)\ u.$$

Apply your result to obtain and solve the transport equation for the modified one-dimensional wave equation when a linear viscous term has been added (namely, a term proportional to the velocity). Are the (projected) characteristics affected by this addition?

References

1. Courant R, Hilbert D (1962) Methods of mathematical physics, vol 2. Interscience, Wiley, New York
2. Garabedian PR (1964) Partial differential equations. Wiley, New York
3. Hadamard J (1903) Leçons sur la Propagation des Ondes et les Équations de l'Hydrodynamique. Hermann, Paris. www.archive.org
4. John F (1982) Partial differential equations. Springer, Berlin
5. Truesdell C, Toupin RA (1960) The classical field theories. In: Flügge S (ed), Handbuch der Physik. Springer, Berlin

Chapter 7
Systems of Equations

Quasi-linear equations of order higher than 1 and systems of quasi-linear equations of any order can be subjected to a classification akin to that of the single second-order equation. Between the two extremes of totally hyperbolic and elliptic types a larger variety of intermediate types can be found. The analysis of totally hyperbolic equations and systems is particularly fruitful as it leads to the concepts of characteristic manifolds and bi-characteristic lines or rays. Physically, the former represent wave fronts propagating through space and the latter are the lines along which signals propagate. As in the case of a single second-order hyperbolic PDE, it is possible to predict the trajectories and variation of amplitude of these signals without necessarily solving the original equations themselves.

7.1 Systems of First-Order Equations

7.1.1 Characteristic Directions

A system of quasi-linear first-order PDEs for n functions $u_1, \ldots, u_n$ of two independent variables x and y can be written as

$$\begin{bmatrix} a_{11} & \ldots & a_{1n} \\ \cdot & \ldots & \cdot \\ \cdot & \ldots & \cdot \\ \cdot & \ldots & \cdot \\ a_{n1} & \ldots & a_{nn} \end{bmatrix} \begin{Bmatrix} u_1 \\ \cdot \\ \cdot \\ \cdot \\ u_n \end{Bmatrix}_{,x} + \begin{bmatrix} b_{11} & \ldots & b_{1n} \\ \cdot & \ldots & \cdot \\ \cdot & \ldots & \cdot \\ \cdot & \ldots & \cdot \\ b_{n1} & \ldots & b_{nn} \end{bmatrix} \begin{Bmatrix} u_1 \\ \cdot \\ \cdot \\ \cdot \\ u_n \end{Bmatrix}_{,y} = \begin{Bmatrix} c_1 \\ \cdot \\ \cdot \\ \cdot \\ c_n \end{Bmatrix}, \tag{7.1}$$

where a_{ij}, b_{ij}, c_i are differentiable functions of x, y and u. We will also use the block matrix notation

M. Epstein, *Partial Differential Equations*, Mathematical Engineering,
DOI 10.1007/978-3-319-55212-5_7

$$\mathbf{A}\mathbf{u}_{,x} + \mathbf{B}\mathbf{u}_{,y} = \mathbf{c}. \tag{7.2}$$

As before, we attempt to find whether or not, given the values $\mathbf{u} = \hat{\mathbf{u}}(r)$ of the vector $\mathbf{u}$ on a curve $x = \hat{x}(r)$, $y = \hat{y}(r)$, we can calculate the derivatives $\mathbf{u}_{,x}$ and $\mathbf{u}_{,y}$ throughout the curve. In complete analogy with Eq. (6.5), we can write

$$\frac{d\hat{\mathbf{u}}}{dr} = \mathbf{u}_{,x}\frac{d\hat{x}}{dr} + \mathbf{u}_{,y}\frac{d\hat{y}}{dr}. \tag{7.3}$$

This vector equation represents, in fact, n scalar equations. Combining this information with that provided by the system of PDEs itself, we obtain at each point of the curve a system of $2n$ algebraic equations, namely,

$$\begin{bmatrix} \frac{d\hat{x}}{dr} & 0 & \cdots & 0 & \frac{d\hat{y}}{dr} & 0 & \cdots & 0 \\ 0 & \frac{d\hat{x}}{dr} & \cdots & 0 & 0 & \frac{d\hat{y}}{dr} & \cdots & 0 \\ . & . & \cdots & . & . & . & \cdots & . \\ . & \cdots & \cdots & . & . & . & \cdots & . \\ . & . & \cdots & . & . & . & \cdots & . \\ 0 & 0 & \cdots & \frac{d\hat{x}}{dr} & 0 & 0 & \cdots & \frac{d\hat{y}}{dr} \\ a_{11} & a_{12} & \cdots & a_{1n} & b_{11} & b_{12} & \cdots & b_{1n} \\ a_{21} & a_{22} & \cdots & a_{2n} & b_{21} & b_{22} & \cdots & b_{2n} \\ . & . & \cdots & . & . & . & \cdots & . \\ . & . & \cdots & . & . & . & \cdots & . \\ . & . & \cdots & . & . & . & \cdots & . \\ a_{n1} & a_{n2} & \cdots & a_{nn} & b_{n1} & b_{n2} & \cdots & b_{nn} \end{bmatrix} \begin{Bmatrix} u_{1,x} \\ u_{2,x} \\ . \\ . \\ . \\ u_{n,x} \\ u_{1,y} \\ u_{2,y} \\ . \\ . \\ . \\ u_{n,y} \end{Bmatrix} = \begin{Bmatrix} \frac{d\hat{u}_1}{dr} \\ \frac{d\hat{u}_2}{dr} \\ . \\ . \\ . \\ \frac{d\hat{u}_n}{dr} \\ c_1 \\ c_2 \\ . \\ . \\ . \\ c_n \end{Bmatrix}, \tag{7.4}$$

which can be written more compactly in terms of partitioned matrices as

$$\begin{bmatrix} \frac{d\hat{x}}{dr}\mathbf{I} & \frac{d\hat{y}}{dr}\mathbf{I} \\ \\ \mathbf{A} & \mathbf{B} \end{bmatrix} \begin{Bmatrix} \mathbf{u}_{,x} \\ \\ \mathbf{u}_{,y} \end{Bmatrix} = \begin{Bmatrix} \frac{d\hat{\mathbf{u}}}{dr} \\ \\ \mathbf{c} \end{Bmatrix}, \tag{7.5}$$

where $\mathbf{I}$ is the unit matrix of order n. If the determinant of this system of $2n$ linear equations is not zero, we obtain (point-wise along the curve) a unique solution for the local values of the partial derivatives. If the determinant vanishes, we obtain a (projected) characteristic direction. According to a result from linear algebra [2], the determinant of a partitioned matrix of the form

$$\mathbf{R} = \begin{bmatrix} \mathbf{M} & \mathbf{N} \\ \\ \mathbf{P} & \mathbf{Q} \end{bmatrix}, \tag{7.6}$$

where $\mathbf{M}$ and $\mathbf{Q}$ are square sub-matrices, is obtained as

$$\det \mathbf{R} = \det \mathbf{Q} \ \det \left(\mathbf{M} - \mathbf{N}\mathbf{Q}^{-1}\mathbf{P} \right), \tag{7.7}$$

assuming that $\mathbf{Q}$ is non-singular. Accordingly, if we assume that $\mathbf{B}$ in (7.5) is non-singular, we obtain the following condition for a direction to be characteristic

$$\det \left(\frac{d\hat{x}}{dr}\mathbf{I} - \frac{d\hat{y}}{dr}\mathbf{B}^{-1}\mathbf{A} \right) = 0, \tag{7.8}$$

which is the same as the generalized eigenvalue problem for the characteristic slopes dx/dy, that is,

$$\det \left(\mathbf{A} - \frac{dx}{dy}\mathbf{B} \right) = 0. \tag{7.9}$$

The same result is obtained if we assume that $\mathbf{A}$, instead of $\mathbf{B}$, is non-singular.

If the eigenvalues $\lambda = dx/dy$ of the matrix $\mathbf{A}$ weighted by $\mathbf{B}$, as indicated in Eq. (7.9), are all real and distinct, the system is called *totally hyperbolic*. At the other extreme, if n is even and all the eigenvalues are complex, the system is *totally elliptic*. Intermediate cases are also possible. Notice that for quasi-linear systems the type depends also on the solution.

7.1.2 Weak Singularities

Assuming the solution $\mathbf{u}$ to be continuous, we investigate the possibility of propagation of weak signals, that is, those carrying a discontinuity in the first or higher derivatives. Taking jumps of the system (7.2) yields

$$\mathbf{A}[\![\mathbf{u}_{,x}]\!] + \mathbf{B}[\![\mathbf{u}_{,y}]\!] = 0. \tag{7.10}$$

On the other hand, by virtue of the geometric compatibility conditions (6.27), we obtain that, in the direction tangential to a possible line of discontinuity $x = \tilde{x}(s)$, $y = \tilde{y}(s)$,

$$[\![\mathbf{u}_{,x}]\!]\frac{d\tilde{x}}{ds} + [\![\mathbf{u}_{,y}]\!]\frac{d\tilde{y}}{ds} = 0. \tag{7.11}$$

Combining these results we conclude that a weak discontinuity can only occur if

$$\left(\mathbf{A} - \frac{dx}{dy}\mathbf{B} \right) [\![\mathbf{u}_{,x}]\!] = \mathbf{0}. \tag{7.12}$$

In this way, we arrive at the same conclusion as before, namely, that weak discontinuities of the solution can only exist across characteristic curves. Totally elliptic systems cannot sustain any discontinuities.

7.1.3 Strong Singularities in Linear Systems

A *strong singularity* of a solution of a PDE is a discontinuity of order smaller than the order of the PDE. In a system of first-order PDEs, therefore, a strong singularity is a discontinuity of the solution itself. We have already encountered such discontinuities in the formation of shocks out of perfectly smooth initial conditions. These shocks, however, do not propagate along characteristics. On the other hand, for strictly linear equations, where the projected characteristics are independent of the solution, discontinuities can only arise as a consequence of the specification of discontinuous initial data, as we have already learned in Example 3.2. Strong discontinuities in linear hyperbolic systems propagate along characteristics.

Consider the linear system

$$\mathbf{A}\mathbf{u}_{,x} + \mathbf{B}\mathbf{u}_{,y} = \mathbf{C}\mathbf{u} + \mathbf{d}, \tag{7.13}$$

where the $n \times n$ matrices $\mathbf{A}, \mathbf{B}, \mathbf{C}$ and the vector $\mathbf{d}$ are functions of x and y alone. Assume that the eigenvalues $\lambda_1, \ldots, \lambda_n$ obtained from Eq. (7.9) are all real and distinct.[1] Let $\mathbf{M}$ be the *modal matrix*, whose columns are (linearly independent) eigenvectors corresponding to these eigenvalues. Then

$$\mathbf{A}\mathbf{M} = \mathbf{B}\mathbf{M}\boldsymbol{\Lambda}, \tag{7.14}$$

where $\boldsymbol{\Lambda}$ is the diagonal matrix of eigenvalues. Defining a new vector $\mathbf{v}$ of dependent variables by

$$\mathbf{u} = \mathbf{M}\mathbf{v}, \tag{7.15}$$

the system (7.13) can be written in *canonical form* as

$$\boldsymbol{\Lambda}\mathbf{v}_{,x} + \mathbf{v}_{,y} = \hat{\mathbf{C}}\mathbf{v} + \hat{\mathbf{d}}, \tag{7.16}$$

where the new matrix $\hat{\mathbf{C}}$ and the new vector $\hat{\mathbf{d}}$ are still functions of x and y alone. What we have achieved, at almost no cost, is a *decoupling* of the main part of each equation of the system in the sense that each equation contains a derivation in only one of the characteristic directions, as follows directly from the fact that $\lambda_i = dx/dy$ is the local slope of the i-th characteristic line, as suggested in Fig. 7.1. Thus, the dependent variable v_i can attain different values v_i^-, v_i^+ on either side of the i-th characteristic line without violating the i-th differential equation. Moreover, it is also possible to obtain the variation of the amplitude of the jump $[\![v_i]\!]$ along the characteristic line.

---

[1] There may be multiple eigenvalues, as long as the dimension of the corresponding eigenspaces is equal to the multiplicity.

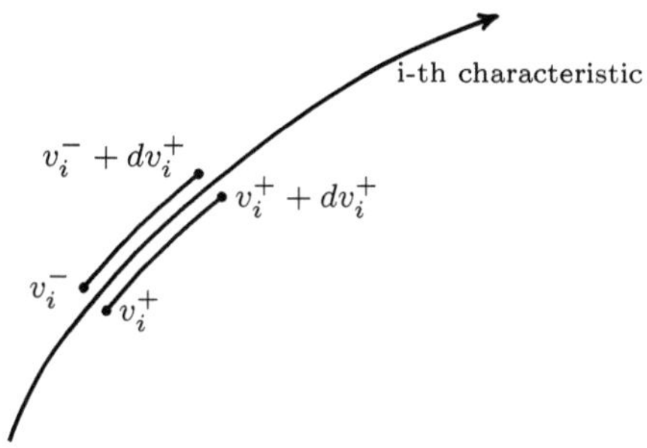

Fig. 7.1 Strong discontinuities in linear systems

7.1.4 *An Application to the Theory of Beams*

7.1.4.1 The Bernoulli–Euler Model

The classical theory of beams, named after Daniel Bernoulli (1700–1782) and Leonhard Euler (1707–1783), is one of the cornerstones of structural engineering. It is based on a number of simplifying assumptions, some of which are the following: (a) the beam is symmetric with respect to a plane; (b) the beam axis is straight; (c) the supports and the loading are symmetrical about the plane of symmetry of the beam; (c) the loading is transversal, that is, perpendicular to the axis; (d) the deflections of the axis are transversal and very small when compared with the beam dimensions; (e) the material abides by Hooke's law of linear elasticity; (f) the plane normal cross sections of the beam remain plane and perpendicular to the deformed axis; (g) the rotary inertia of the cross sections is neglected in the dynamic equations. The last two assumptions are crucial to the simplicity and usefulness of the theory while also embodying some of its limitations.

On the basis of these assumptions, the equations of classical beam theory are not difficult to derive. Introducing x, y Cartesian coordinates in the plane of symmetry and aligning the x axis with the beam axis, as shown in Fig. 7.2, we denote by $q = q(x)$ the transverse load per unit length (positive downwards) and by $w = w(x, t)$ the transverse deflection (positive upwards). The time coordinate is denoted by t. Assuming for specificity a constant cross section of area A and centroidal moment of inertia I, the governing equations can be written as

$$\left\{\begin{array}{lcc} V_x & = & -q - \rho A w_{tt} \\ M_x & = & V \\ EI\theta_x & = & M \\ w_x & = & \theta \end{array}\right. \tag{7.17}$$

In these equations ρ denotes the constant density, E is the modulus of elasticity, $\theta = \theta(x, t)$ is the rotation of the cross section, and $V = V(x, t)$ and $M = M(x, t)$ are, respectively the internal shear force and bending moment resulting from the axial and transversal stresses in the beam cross section. Notice that the assumption

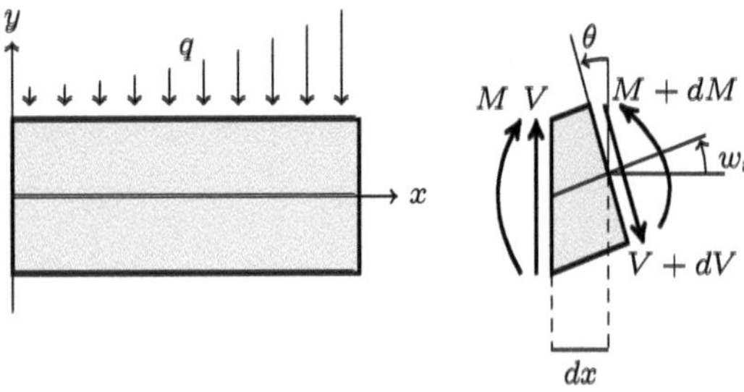

Fig. 7.2 Beam theory derivation

of perpendicularity between the cross sections and the deformed axis implies the vanishing of the corresponding shear strains, but not of the shear stresses, an internal contradiction of the theory. The first two equations are the result of enforcing vertical and rotational dynamic equilibrium, while the third equation arises from Hooke's law. The fourth equation establishes the perpendicularity condition by equating the slope of the axis to the rotation of the cross section.

Introducing the linear and angular velocities $v = w_t$ and $\omega = \theta_t$, respectively, the system (7.17) can be rewritten as the first-order system

$$\begin{cases} V_x &= -q - \rho A v_t \\ M_x &= V \\ EI\omega_x &= M_t \\ v_x &= \omega \end{cases} \tag{7.18}$$

This system can be written as

$$\mathbf{A}\mathbf{u}_{,t} + \mathbf{I}\mathbf{u}_{,x} = \mathbf{c}, \tag{7.19}$$

where

$$\mathbf{A} = \begin{bmatrix} 0 & 0 & 0 & \rho A \\ 0 & 0 & 0 & 0 \\ 0 & -\frac{1}{EI} & 0 & 0 \\ 0 & 0 & 0 & 0 \end{bmatrix} \qquad \mathbf{u} = \begin{Bmatrix} V \\ M \\ \omega \\ v \end{Bmatrix} \qquad \mathbf{c} = \begin{Bmatrix} -q \\ V \\ 0 \\ \omega \end{Bmatrix}. \tag{7.20}$$

Solving the eigenvalue problem

$$\det\left(\mathbf{A} - \frac{dt}{dx}\mathbf{I}\right) = 0, \tag{7.21}$$

we obtain the trivial condition

$$\left(\frac{dt}{dx}\right)^4 = 0. \tag{7.22}$$

This means that any disturbance would travel at infinite speed.

7.1.4.2 The Timoshenko Beam

To remedy the deficiencies of the classical theory, Stephen Timoshenko (1887–1972) proposed in the early 1920s a modified theory that bears his name. He made just two modifications to the theory, as follows: (a) plane normal cross sections remain plane but do not necessarily remain perpendicular to the deformed axis; (b) the rotational inertia of the cross sections is included in the dynamic equilibrium equations. As a result, the rigid (positive counterclockwise) rotation θ of the cross sections is no longer equal to the slope of the axis and a shear strain equal to their difference is related, via Hooke's shear elastic modulus G, to the shear force. Since the shear strain is constant at each section (rather than quadratic) an equivalent "shear area" A_s is introduced to compensate for this deliberate error in the theory. The equations of the Timoshenko beam theory result in the system (7.19) but with the new coefficient matrix

$$\mathbf{A} = \begin{bmatrix} 0 & 0 & 0 & \rho A \\ 0 & 0 & -\rho I & 0 \\ 0 & -\frac{1}{EI} & 0 & 0 \\ \frac{1}{GA_s} & 0 & 0 & 0 \end{bmatrix}. \tag{7.23}$$

The eigenvalue problem yields now four distinct roots

$$\left(\frac{dt}{dx}\right)_{1,2} = \pm\sqrt{\frac{\rho}{E}} \qquad \left(\frac{dt}{dx}\right)_{3,4} = \pm\sqrt{\frac{\rho A}{GA_s}}. \tag{7.24}$$

The first two roots are the inverse of the speeds of propagation of the 'bending waves', while the last two roots are the inverse of the speeds of propagation of the 'shear waves'. The importance of these finite speeds in aircraft design was first recognized in [5], where a clear exposition of the evolution of strong discontinuities in beams is presented from basic principles and extended to the case of multiple eigenvalues. Notice how, rather than solving specific numerical problems, a careful analysis of the structure of the equations has allowed us to arrive at significant conclusions about the phenomena at hand with relatively elementary computations.

7.1.5 Systems with Several Independent Variables

In the case of n quasi-linear first-order equations for n functions $u_i = u_i(x_1, \dots, x_K)$, with $i = 1, \dots, n$, of K independent variables, we write the system in matrix notation as

$$\mathbf{A}_1\mathbf{u}_{,x_1} + \mathbf{A}_2\mathbf{u}_{,x_2} + \cdots + \mathbf{A}_K\mathbf{u}_{,x_K} = \mathbf{c}, \tag{7.25}$$

or, more compactly, using the summation convention of Box 2.2 for repeated capital indices in the range $1, \dots, K$,

$$\mathbf{A}_I \mathbf{u}_{,x_I} = \mathbf{c}. \tag{7.26}$$

The various matrices $\mathbf{A}_I$ and the vector $\mathbf{c}$ are assumed to be differentiable functions of the arguments $x_1, \ldots, x_K, u_1, \ldots, u_n$. We want to search for *characteristic manifolds*. As usual, these are special $(K-1)$-dimensional hyper-surfaces in the space of independent variables. On these characteristic manifolds, the specification of initial data is not sufficient to guarantee the existence of a unique solution in a neighbourhood of the manifold. Alternatively, these manifolds can be regarded as carriers of weak singularities, that is, discontinuities in the first (or higher) derivatives of the solutions. In Sect. 6.4.1 we obtained an important result to the effect that if a function is continuous across a singular surface the jump of its gradient is necessarily collinear with the normal $\mathbf{n}$ (in the usual Cartesian metric) to the surface. Thus, we may write

$$[\![\mathbf{u}_{,x_I}]\!] = \mathbf{a}\, n_I \qquad I = 1, \ldots, K, \tag{7.27}$$

where $\mathbf{a}$ is a vector of n entries.

Taking the jump of Eq. (7.26) and invoking the last result yields

$$(n_I\ \mathbf{A}_I)\mathbf{a} = \mathbf{0}. \tag{7.28}$$

For the vector $\mathbf{a}$, containing the intensity of the jumps, to be non-zero the determinant of the coefficient matrix must vanish, that is,

$$\det(n_I\ \mathbf{A}_I) = 0. \tag{7.29}$$

This is the equation defining the possible local normals to characteristic manifolds. We ask ourselves: what kind of equation is this? It is obviously a homogeneous polynomial of degree n in the K components n_I. It may so happen that no vector $\mathbf{n}$ (that is, no direction) satisfies this equation, in which case the system is said to be *totally elliptic* at the point in question. At the other extreme, if fixing $K-1$ entries of $\mathbf{n}$ the polynomial in the remaining component has n distinct real roots, we have a case of a *totally hyperbolic* system at the point. Let a putative singular hyper-surface be given by an equation such as

$$\phi(x_1, \ldots, x_K) = 0. \tag{7.30}$$

Then the vector $\mathbf{n}$, not necessarily of unit length, can be identified with the gradient of ϕ, namely,

$$n_I = \phi_{,x_I} \tag{7.31}$$

so that we get

$$\det(\mathbf{A}_I\ \phi_{,x_I}) = 0. \tag{7.32}$$

This is a highly non-linear single first-order PDE for the function ϕ. We call it the *characteristic equation* or the *generalized eikonal equation* associated with the system. A characteristic manifold (hyper-surface) is the level curve $\phi = 0$ of a solution ϕ of this eikonal equation. An interesting idea consists of investigating the characteristics of the eikonal equation. They are called the *bi-characteristics* of the original system. These lines play a fundamental role in the theory of wave propagation and rays. In many applications, it is convenient to single out one of the independent variables as the time coordinate, say $x_K = t$. The physical meaning of a singular surface $\phi(x_1, \dots, x_{K-1}, t) = 0$ in space-time is a propagating *wave front*. If we write the equation of the singular surface as

$$\psi(x_1, \dots, x_{K-1}) - t = 0, \tag{7.33}$$

the *spatial wave front* at any instant of time t_0 is the level curve $\psi = t_0$, as suggested in Fig. 7.3 for the case $K = 3$. This observation leads to a somewhat friendlier formulation of the characteristic equation, as described in Box 7.1.

Box 7.1 A friendlier formulation of the characteristic equation
We already had an opportunity to remark that the solutions ϕ of Eq. (7.32) do not directly represent a characteristic hyper-surface. Indeed, we need to impose the additional constraint $\phi = 0$. This feature can be circumvented by assuming, as is often the case in applications, that a time variable $t = x_K$ can be singled out and that the derivative of ϕ with respect to t does not vanish. Introducing the new representation $\phi(x_, \dots, x_{K-1}, t) = \psi(x_1, \dots, x_{K-1}) - t$, the characteristic equation (7.32) becomes

$$\det(\mathbf{A}_\alpha \, \phi_{,x_\alpha} - \mathbf{A}_K) = 0,$$

where the summation convention for Greek indices is restricted to the range $1, \dots, K-1$. Every solution of this first-order non-linear PDE represents a characteristic manifold. In terms of the *normal speed* of a moving surface, introduced in Box 7.2, this equation can be written point-wise as the algebraic condition

$$\det(\mathbf{A}_\alpha \, m_\alpha - V \mathbf{A}_K) = 0,$$

in which m_α are the component of the unit normal to the spatial wave front. In terms of classification, the system is *totally elliptic* if there are no real eigenvalues V. It is *totally hyperbolic* if it has $K-1$ distinct real eigenvalues. Repeated eigenvalues can also be included in the definition, since they appear in applications.

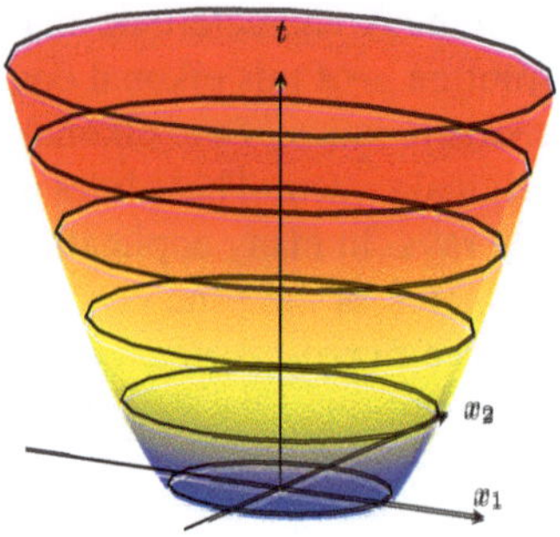

Fig. 7.3 Spatial wave fronts as level sets of a spatiotemporal wave front

7.2 Systems of Second-Order Equations

7.2.1 Characteristic Manifolds

Although it is possible to express any higher-order equation or system thereof by means of a system of first-order equations, there is often a loss of physical meaning in the process of introducing artificial intermediate variables and equations. For this reason, it is convenient in many cases to leave the equations in their original form, such as obtained in applications in Solid and Fluid Mechanics and many other fields in Engineering and Physics. A system of n second-order quasi-linear equations for n functions of K variables can be written compactly as

$$\mathbf{A}_{IJ}\mathbf{u}_{,x_I x_J} = \mathbf{c}, \tag{7.34}$$

where the matrices $\mathbf{A}_{IJ}$ and the vector $\mathbf{c}$ are assumed to be differentiable functions of the independent variables x_I and possibly also of the unknown functions u_i and their first derivatives $u_{i,I}$. We proceed to evaluate the jump of this equation under the assumption that neither the functions u_i nor their first derivatives undergo any discontinuities, since we are looking for weak discontinuities. The result is

$$\mathbf{A}_{IJ}[\![\mathbf{u}_{,x_I x_J}]\!] = \mathbf{0}. \tag{7.35}$$

Invoking the iterated compatibility condition derived in Box 6.1, we can write

$$\mathbf{A}_{IJ}\ \mathbf{a}\ n_I n_J = \mathbf{0}, \tag{7.36}$$

where $\mathbf{a}$ is a vector with n entries known as the *wave amplitude* vector. We conclude that a hyper-surface element with normal $\mathbf{n}$ is characteristic if

$$\det\left(\mathbf{A}_{IJ}\ n_I n_J\right) = 0. \tag{7.37}$$

The classification of the system as totally hyperbolic, elliptic or otherwise follows in the usual way. In applications to Elasticity, the purely spatial part of the tensor $\mathbf{A}_{IJ}\, n_I n_J$ is known as the *acoustic tensor* associated with a given direction in space.

Box 7.2 Moving surfaces in $\mathbb{R}^3$

Consider a differentiable function $f(x_1, x_2, x_3, t)$ of the spatial Cartesian coordinates x_1, x_2, x_3 and time t. The equation

$$f(x_1, x_2, x_3, t) = 0$$

represents a fixed three-dimensional hyper-surface in $\mathbb{R}^4$. There is an alternative, more physical, way to view this equation. Indeed, for each value of t we obtain an ordinary two-dimensional surface in the space with coordinates x_1, x_2, x_3. Accordingly, we can regard the equation $f(x_1, x_2, x_3, t) = 0$ as a one-parameter family of surfaces $\mathcal{S}_t$ in ordinary space. Since the parameter is time, we can regard this family as a *moving surface* in space.

An important question is: what is the velocity of this moving surface? Clearly, the velocity is a point-wise notion. Each point of the surface has its own velocity, and this velocity, in general, varies with time. But, what is a 'point' of this surface? Assume that we were to attach, at time t, an observer to a point P moving with the surface. Where would this observer be located at time $t + dt$? There is an infinite number of ways to establish a one-to-one correspondence between points in $\mathcal{S}_t$ and $\mathcal{S}_{t+dt}$! Choosing a particular point P' in $\mathcal{S}_{t+dt}$ we obtain a vector with components dx_1, dx_2, dx_3. These increments are constrained by the vanishing of f on all surfaces of the family by the equation

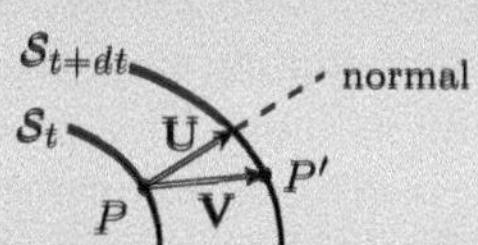

$$df = f_{,i}dx_i + f_t dt = 0,$$

where the summation convention is used for the spatial indices. Dividing by dt and recognizing the vector with components $V_i = dx_i/dt$ as the velocity associated with the pairing P, P' yields

$$V_i\, f_{,i} = -f_t \qquad \text{or} \qquad \mathbf{V} \cdot \nabla f = -f_t.$$

The gradient ∇f is proportional to the unit normal $\mathbf{n}$ to the surface at P, the constant of proportionality being the magnitude of ∇f. We conclude that

$$U = V_n = \mathbf{V} \cdot \mathbf{n} = -\frac{f_t}{\sqrt{\nabla f \cdot \nabla f}}.$$

The right-hand side of this equation is clearly independent of the particular P' chosen as a pair of P. We conclude that: *The normal speed v_n of the moving*

surface at P is a meaningful intrinsic geometric quantity. We call the vector $\mathbf{U} = U\,\mathbf{n}$ the *normal velocity* of the moving surface at P.
If the surface is moving on a material background, the material particle instantaneously occupying the position P at time t has a velocity $\mathbf{v}$. The difference $U_p = U - \mathbf{v} \cdot \mathbf{n}$ is the relative speed of the surface with respect to the particle, also called the *speed of propagation*. If it vanishes identically, we say that the moving surface is *material* since it is dragged by the motion and is always occupied by the same material particles.

7.2.2 Variation of the Wave Amplitude

7.2.2.1 Jump Relations

A peculiar feature of hyperbolic equations and systems is that, without necessarily solving the field equations, it is possible to obtain a wealth of information about the solution. In particular, when the characteristic manifolds are obtained and interpreted as wave fronts, it becomes possible to describe events taking place at those wave fronts, such as the decay or growth of the amplitude of an initially imposed discontinuity. The equation governing this growth or decay is an ordinary differential equation known as the *transport equation* or the *decay-induction equation*. We have already encountered this equation in Sect. 6.4.3 for the general quasi-linear second-order equation in two independent variables. The situation is somewhat more involved when the number of independent variables is larger than 2. The technique used in deriving the transport equation consists of differentiating the PDE with respect to one of the variables, t say, and then calculating its jump. This procedure seems to complicate matters, since the jumps of third derivatives appear in the equation. It can be shown, however, that these derivatives are affected by the same matrix coefficient as the wave amplitude vector in Eq. (7.36). This circumstance enables the elimination of this term by projecting the equation over the wave amplitude corresponding to the particular wave front being dealt with. A general treatment of this issue is presented in very clear terms in [7] for quasi-linear systems of first-order equations. An application to a particular system of second-order equations is given in [1]. Here we present a more or less complete treatment for a particular linear system of second-order equations. The procedure can be extended to the general quasi-linear system by more or less obvious modifications.

An ingenious device is exploited in [7] which allows to easily construct a new set of independent variables, $K - 1$ of which lie in the space-time wave front. Assuming this wave front to have been found and expressed in the form (7.33), a change of variables is introduced as

$$\begin{cases} \xi_I = x_I \quad \text{for } I = 1, \ldots, K-1 \\ \tau = \psi(x_1, \ldots, x_{K-1}) - t \end{cases} \tag{7.38}$$

It follows that for $\tau = 0$ the coordinate lines corresponding to the new variables ξ_I lie on the singular surface. Let $g = g(x_1, \ldots, x_{K-1}, t)$ be a differentiable function of the old variables. It can be readily converted into a function $\hat{g}$ of the new variables by the composition

$$g(x_1, \ldots, t) = g(\xi_1, \ldots, \psi(\xi_1, \ldots) - \tau) = \hat{g}(\xi_1, \ldots, \tau). \tag{7.39}$$

For the sake of compactness, let us abuse the notation and denote by commas the partial derivatives with respect to either x_I or ξ_I. The distinction will be clear from the name of the function (un-hatted or hatted, respectively). In the same vein, let us denote by superimposed dots partial derivatives with respect to t or τ. With this understanding, we obtain

$$\begin{cases} \hat{g}_{,I} = g_I + \dot{g}\,\psi_{,I} \quad \text{for } I = 1, \ldots, K-1 \\ \dot{\hat{g}} = -\,\dot{g} \end{cases} \tag{7.40}$$

The beauty of the new variables is that the coordinates ξ_I are *interior coordinates*, as a result of which *derivatives in those* $K - 1$ *directions do not experience any jump*! Consequently, ξ-derivatives commute with the jump operator. Since we are dealing with weak waves, jumps of functions of the variables ξ_I, τ will occur only in quantities with two or more τ derivatives. Differentiating Eq. (7.40), we obtain the following relations

$$[\![\ddot{g}]\!] = [\![\ddot{\hat{g}}]\!] \tag{7.41}$$
$$[\![\dot{g}_{,I}]\!] = -[\![\ddot{\hat{g}}]\!]\,\psi_{,I} \tag{7.42}$$
$$[\![g_{,IJ}]\!] = [\![\ddot{\hat{g}}]\!]\,\psi_{,I}\psi_{,J} \tag{7.43}$$
$$[\![\dddot{g}]\!] = -[\![\dddot{\hat{g}}]\!] \tag{7.44}$$
$$[\![\ddot{g}_{,I}]\!] = [\![\ddot{\hat{g}}]\!]\,\psi_{,I} + [\![\dddot{\hat{g}}]\!]\psi_{,I} \tag{7.45}$$
$$[\![\dot{g}_{,IJ}]\!] = -[\![\ddot{\hat{g}}]\!]_{,I}\psi_{,J} - [\![\ddot{\hat{g}}]\!]_{,J}\psi_{,I} - [\![\ddot{\hat{g}}]\!]\psi_{,IJ} - [\![\dddot{\hat{g}}]\!]\psi_{,I}\psi_{,J}. \tag{7.46}$$

7.2.2.2 The Transport Equation

For definiteness, we consider a linear system of n equations for n functions $u_1, \ldots, u_n$ of K variables $x_1, \ldots, x_{K-1}, t$ in the normal form

$$\mathbf{A}_{IJ}\,\mathbf{u}_{,IJ} = \ddot{\mathbf{u}}. \tag{7.47}$$

The matrices $\mathbf{A}_{IJ}$ are assumed to be constant and symmetric. Taking jumps and invoking Eqs. (7.41) and (7.43) yields

$$\left(\mathbf{A}_{IJ}\,\psi_{,I}\psi_{,J} - \mathbf{I}\right)\,[\![\ddot{\hat{\mathbf{u}}}]\!] = \mathbf{0}, \tag{7.48}$$

in which $\mathbf{I}$ is the unit matrix of order n. We assume the system to be totally hyperbolic. A solution $\psi = \psi(x_1, \ldots, x_{K-1})$ of the first-order eikonal equation

$$\det\left(\mathbf{A}_{IJ}\,\psi_{,I}\psi_{,J} - \mathbf{I}\right) = 0 \tag{7.49}$$

provides us with a spatiotemporal wave front $\psi(x_1, \ldots, x_{K-1}) - t = 0$. At each point of this hyper-surface, we obtain a single eigen-direction satisfying Eq. (7.48).

If we differentiate the original PDE with respect to time and then take the jumps while invoking Eqs. (7.41) and (7.46), we can write

$$\mathbf{A}_{IJ}\left(-[\![\ddot{\hat{u}}]\!]_{,I}\psi_{,J} - [\![\ddot{\hat{u}}]\!]_{,J}\psi_{,I} - [\![\ddot{\hat{u}}]\!]\psi_{,IJ} - [\![\dddot{\hat{u}}]\!]\psi_{,I}\psi_{,J}\right) = -[\![\dddot{\hat{\mathbf{u}}}]\!], \tag{7.50}$$

or

$$\mathbf{A}_{IJ}\left([\![\ddot{\hat{u}}]\!]_{,I}\psi_{,J} + [\![\ddot{\hat{u}}]\!]_{,J}\psi_{,I} + [\![\ddot{\hat{u}}]\!]\psi_{,IJ}\right) + \left(\mathbf{A}_{IJ}\psi_{,I}\psi_{,J} - \mathbf{I}\right)[\![\dddot{\hat{u}}]\!] = \mathbf{0}. \tag{7.51}$$

Since the matrices $\mathbf{A}_{IJ}$ are symmetric, the right and left eigenvectors of $\mathbf{A}_{IJ}\psi_{,I}\psi_{,J} - \mathbf{I}$ are the same. On multiplying to the left by the transpose of $[\![\ddot{\hat{u}}]\!]$ we cancel out the term containing the jump of the third derivative. The result is

$$[\![\ddot{\hat{u}}]\!]^T\,\mathbf{A}_{IJ}\left([\![\ddot{\hat{u}}]\!]_{,I}\psi_{,J} + [\![\ddot{\hat{u}}]\!]_{,J}\psi_{,I} + [\![\ddot{\hat{u}}]\!]\psi_{,IJ}\right) = 0. \tag{7.52}$$

Recall that the eigenvector is known in direction and that, therefore, this equation becomes an ODE for the amplitude a of $[\![\ddot{\hat{u}}]\!]$. This is the desired transport equation. It can be shown [1] that its characteristics coincide with the bi-characteristics of the original equation.

7.2.3 *The Timoshenko Beam Revisited*

7.2.3.1 Characteristics

We have derived in Sect. 7.1.4 the dynamic equations for a Timoshenko beam as a system of 4 first-order PDEs. It is also possible to express these equations in terms of 2 second-order PDEs for the displacement w and the rotation θ, respectively. The result[2] for a beam of homogeneous properties is

$$\begin{cases} \rho A w_{tt} - GA_s(w_x - \theta)_x = -q \\ \rho I \theta_{tt} - EI\theta_{xx} = GA_s(w_x - \theta) \end{cases} \tag{7.53}$$

[2] See Exercise 7.5.

These equations can be recast in the normal form (7.47), except for the addition of a vector **c** on the right-hand side. Since $K = 2$ we obtain the single matrix

$$\mathbf{A} = \begin{bmatrix} \frac{GA_s}{\rho A} & 0 \\ 0 & \frac{E}{\rho} \end{bmatrix} \tag{7.54}$$

The additional term **c** is given by

$$\mathbf{c} = \begin{Bmatrix} \frac{q+GA_s}{\rho A}\theta_x \\ -\frac{GA_s}{\rho I}(w_x - \theta) \end{Bmatrix} \tag{7.55}$$

Equation (7.49) provides us in this case immediately with the eigenvalues

$$\left(\frac{dt}{dx}\right)_1^2 = \frac{\rho A}{GA_s} \qquad \left(\frac{dt}{dx}\right)_2^2 = \frac{\rho}{E}. \tag{7.56}$$

which, except for a different numbering, are identical to those obtained earlier as Eq. (7.24). The corresponding eigenvectors are, respectively, $\langle 1, 0\rangle$ and $\langle 0, 1\rangle$. Thus, the characteristics are straight lines. We have four families of characteristics in the x, t plane, each pair with slopes of different signs. For total hyperbolicity, these slopes are all distinct. The interesting case of double eigenvalues is considered separately in Box 7.3.

7.2.3.2 Variation of Amplitude

The general prescription of Eq. (7.52) has to be modified slightly to take into consideration the additional vector **c**. Indeed, this term contributes first partial derivatives, which have been assumed to be continuous and, accordingly, do not affect the characteristic equations. On the other hand, when constructing the transport equation, a further derivative is introduced which affects the formation of the jumps. The additional term is

$$[\![\mathbf{c}_t]\!] = \begin{Bmatrix} \frac{GA_s}{\rho A}[\![\theta_{xt}]\!] \\ -\frac{GA_s}{\rho I}[\![w_{xt}]\!] \end{Bmatrix} \tag{7.57}$$

The final result corresponding to Eq. (7.52) for the first eigenvalue is

$$\langle 1\; 0\rangle\left(\begin{bmatrix} \frac{GA_s}{\rho A} & 0 \\ 0 & \frac{E}{\rho} \end{bmatrix}\begin{Bmatrix} 2a_x\sqrt{\frac{\rho A}{GA_s}} \\ 0 \end{Bmatrix} + a\sqrt{\frac{GA_s}{\rho A}}\begin{Bmatrix} 0 \\ 1 \end{Bmatrix}\right) = 0. \tag{7.58}$$

This equation implies that $a_x = 0$, that is, the amplitude of a jump in the second derivative of the displacement remains constant as it propagates. A similar analysis for the second eigenvalue shows that the amplitude of the second derivative of the rotation also remains constant. The main reason for these facts, apart from the constancy of the sectional properties, is the complete decoupling between the modes. This decoupling stems from the particular form of the additional term **c** and from the assumption of total hyperbolicity, namely, that the speeds of propagation of the bending and shear signals are different. A surprising discrepancy is found when these speeds happen to be equal to each other, as discussed in Box 7.3.

Box 7.3 When two eigenvalues meet

The shear area A_s of the Timoshenko beam is a matter of some controversy. At any rate, it is an adjustable quantity that may depend on the application at hand [3]. In exceptional cases it may attain a value such that $GA_s = EA$, which would lead to the equality of the eigenvalues of the matrix **A**. In that case, *any vector* $\langle a, b\rangle$ *in the space of jumps with coordinates* $[\![w_{tt}]\!]$, $[\![\theta_{tt}]\!]$ *is an eigenvector*. Equation (7.58) must then be replaced by

$$\begin{bmatrix} \frac{E}{\rho} & 0 \\ 0 & \frac{E}{\rho} \end{bmatrix}\begin{Bmatrix} 2a_x \\ 2b_x \end{Bmatrix} + \frac{E}{\rho}\begin{Bmatrix} b \\ -\frac{A}{I}a \end{Bmatrix} = 0,$$

which leads to the system

$$\begin{cases} 2a_x + b = 0 \\ 2b_x - \frac{A}{I}\, a = 0. \end{cases}$$

In obtaining these equations, a very important observation is that, unlike the case of simple eigenvalues, the projection on the corresponding eigenvector (that is, the multiplication to the left by $[\![\dddot{u}]\!]^T$ in Eq. (7.52)) is not called for, since the term $\left(\mathbf{A}_{IJ}\psi_{,I}\psi_{,J} - \mathbf{I}\right)[\![\hat{u}]\!]$ in Eq. (7.51) cancels out identically in the case of a double eigenvalue. It is this fact that permits us to obtain two equations rather than just one, as was the case of simple eigenvalues. The solution of this coupled system is

$$a = C\cos\frac{x}{2\kappa} + D\sin\frac{x}{2\kappa} \qquad b = -\frac{D}{\kappa}\cos\frac{x}{2\kappa} + \frac{C}{\kappa}\sin\frac{x}{2\kappa},$$

where $\kappa = \sqrt{A/I}$ is the radius of gyration and C, D are constants to be adjusted according to the initial conditions, assuming that this knowledge is available at some point on one of the characteristic lines. The solution has been parametrized by x, but could as well be parametrized by t or any other parameter along the characteristic line. The main conclusion is that, unlike the case of simple eigenvalues, whereby the amplitude does not decay or grow, in the case of a double eigenvalue the amplitudes of both bending and shear signals are coupled and evolve harmonically in time.

7.2.4 *Air Acoustics*

7.2.4.1 Wave Fronts and Rays

A relatively simple example is provided by air acoustics, which is governed by the three-dimensional wave equation

$$p_{xx} + p_{yy} + p_{zz} = \frac{1}{c^2} p_{tt}, \tag{7.59}$$

in which $p = p(x, y, z, t)$ is the *acoustic pressure* and c is the speed of sound.

Writing the equation of the characteristic manifold in the form

$$\phi(x, y, z, t) = \psi(x, y, z) - t = 0, \tag{7.60}$$

as suggested in Box 7.1, we obtain

$$\psi_x^2 + \psi_y^2 + \psi_z^2 - \frac{1}{c^2} = 0. \tag{7.61}$$

Assuming, for simplicity, a two-dimensional situation $\psi = \psi(x, y)$, we can solve the characteristic equation by the method of characteristic strips developed in Chap. 5. We adopt as the initial condition a parabolic wave front given parametrically by

$$x = r \qquad y = r^2 \qquad \psi = 0. \tag{7.62}$$

Physically, this initial wave front could have been produced by a distribution of many speakers or other sound producing devices arranged on a parabolic surface and activated simultaneously at time $t = 0$. Extending these data by the strip condition and by the PDE itself yields

$$\psi_x + 2\psi_y r = 0 \qquad \psi_x^2 + \psi_y^2 - \frac{1}{c^2} = 0. \tag{7.63}$$

The characteristic strips are solutions of the system

$$\frac{dx}{ds} = 2\psi_x \qquad \frac{dy}{ds} = 2\psi_y \qquad \frac{d\psi}{ds} = 2\psi_x^2 + 2\psi_y^2 = \frac{2}{c^2} \qquad \frac{d\psi_x}{ds} = 0 \qquad \frac{d\psi_y}{ds} = 0. \tag{7.64}$$

The solution of this system is given by

$$x = -\frac{2rs}{c\sqrt{1+4r^2}} + r \qquad y = \frac{s}{c\sqrt{1+4r^2}} + r^2 \qquad \psi = \frac{s}{c^2}. \tag{7.65}$$

For any value of s the solutions projected on the x, y plane are straight lines. They are the bi-characteristics of the original PDE. For this particular equation, the bi-characteristics are perpendicular to the spatial wave fronts. On the concave side of the initial parabola these bi-characteristics (or rays) tend to converge. This phenomenon, observed also in optics, is known as focusing and the formation of a caustic. Figure 7.4 provides a plot of the bi-characteristics for our example.

For a circular (cylindrical) surface of radius R and centre at $(0, R)$, all the rays converge to the centre. The initial wave front can be written parametrically as

$$x = R\sin\theta \qquad y = R(1-\cos\theta) \qquad \psi = 0. \tag{7.66}$$

The strip condition (5.27) yields

$$\psi_x R\cos\theta + \psi_y R\sin\theta = 0. \tag{7.67}$$

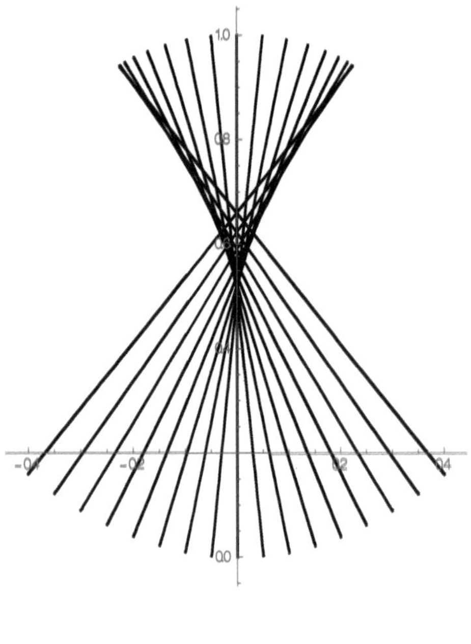

Fig. 7.4 Acoustic focusing

The characteristic ODEs are given again by (7.64). The solution in this case is given by

$$x = -\frac{2s}{c}\sin\theta + R\cos\theta \qquad y = \frac{2s}{c}\cos\theta + R(1-\cos\theta) \qquad \psi = \frac{2s}{c^2}. \tag{7.68}$$

The parameters s, θ are readily eliminated and we obtain the solution of the wave fronts in the form

$$\psi = \frac{1}{c}\left(R - \sqrt{x^2 + (R-y)^2}\right). \tag{7.69}$$

7.2.4.2 Transport Equation

We have a single equation, so that the matrices $\mathbf{A}_{IJ}$ become scalars. In our case, moreover, they attain the simple form $c^2\delta_{IJ}$, where δ_{IJ} is the Kronecker symbol. Correspondingly, the amplitude $[\![\ddot{\mathbf{u}}]\!]$ is just a scalar a and the decay equation (7.52) becomes

$$2a_x\psi_x + 2a_y\psi_y + a(\psi_{xx} + \psi_{yy}) = 0. \tag{7.70}$$

Notice that at this stage the coefficients ψ_x, ψ_y are known. The characteristics of this first-order PDE for a are given by

$$\frac{dx}{dr} = 2\psi_x \qquad \frac{dy}{dr} = 2\psi_y \qquad \frac{da}{dr} = -a(\psi_{xx} + \psi_{yy}). \tag{7.71}$$

As expected, the projected characteristics of the transport equation are precisely the same as the bi-characteristics of the original equation. Thus, weak discontinuities travel along rays (in this example perpendicular to the moving wave front). Let us integrate the transport equation for the circular case, since the parabolic case leads to more involved formulas. Our objective is to satisfy our intuition that as the signals converge toward a focus the acoustic pressure increases without bound. Introducing the solution ψ into (7.71), the system of equations is written as

$$\frac{dx}{dr} = -\frac{2x}{c\rho} \qquad \frac{dy}{dr} = \frac{2(R-y)}{c\rho} \qquad \frac{da}{dr} = \frac{a}{c\rho}, \tag{7.72}$$

where $\rho = \sqrt{x^2 + (R-y)^2}$ is the distance to the centre of the circle. From the first two equations we conclude that

$$\frac{dx}{dy} = \frac{-x}{R-y}, \tag{7.73}$$

so that, as expected, the rays are the radii

$$x = k(R-y). \tag{7.74}$$

Notice that

$$d\rho = \rho_x dx + \rho_y dy = -\frac{2}{c} dr. \tag{7.75}$$

It follows that

$$\frac{da}{d\rho} = \frac{da}{dr}\frac{dr}{d\rho} = -\frac{a}{2\rho}, \tag{7.76}$$

which integrates to

$$a = a_0 \sqrt{\frac{R}{\rho}}, \tag{7.77}$$

where a_0 is the intensity of the excitation (produced, say, by a sudden pulse of the speakers). Thus, the pressure grows without bound as ρ approaches the center of the circle. It is important to realize that we have not solved the acoustic differential equation (7.60). We have merely found the rays as carriers of weak discontinuities and the variation of their amplitude. This information was gathered by solving first-order differential equations only.

7.2.5 *Elastic Waves*

7.2.5.1 Basic Equations of Linear Infinitesimal Elasticity

In Chap. 2 we presented the basic balance equations of Continuum Mechanics. In particular, the equation of balance of linear momentum (2.55) reads

$$\rho \frac{\partial v_i}{\partial t} + \rho v_{i,j} v_j = b_i + \sigma_{ij,j}. \tag{7.78}$$

Here, $\sigma_{ij} = \sigma_{ji}$ are the (Cartesian) components of the stress tensor, ρ is the current density, b_i are the components of the spatial body force and v_i are the components of the velocity vector. This formulation is strictly *Eulerian* or *spatial*. For elastic solids, it is convenient to introduce an alternative (Lagrangian) formulation based on a fixed reference configuration. Nevertheless, for our current purpose, we will adopt a linearized formulation around the present configuration, in which case the distinction between the two formulations alluded to above can be disregarded. Moreover, the inertia term appearing on the left-hand side of Eq. (7.78) will be approximated by the product $\rho \frac{\partial v_i}{\partial t}$, where ρ is identified with the density in the fixed (current) state, so that the mass conservation (continuity equation) does not need to be enforced. Finally, our main kinematic variable is a *displacement vector* field with components $u_i = u_i(x_1, x_2, x_3, t)$ in the global inertial frame of reference. In terms of these variables, the velocity components are given by

$$v_i = \frac{\partial u_i}{\partial t}. \tag{7.79}$$

The material properties will be assumed to abide by *Hooke's law*

$$\sigma_{ij} = \frac{1}{2} C_{ijkl} \left(u_{k,l} + u_{l,k} \right). \tag{7.80}$$

The fourth-order tensor of elastic constants **C** enjoys the symmetries

$$C_{ijkl} = C_{klij} = C_{jikl}. \tag{7.81}$$

It has, therefore, a maximum of 21 independent components. Particular material models enjoy further symmetries, the most symmetric of which is the *isotropic material*, completely characterized by two elastic constants. For the isotropic material we have

$$C_{ijkl} = \lambda \delta_{ij} \delta_{kl} + \mu (\delta_{ik} \delta_{jl} + \delta_{il} \delta_{jk}). \tag{7.82}$$

In this equation λ and μ are the *Lamé coefficients* and δ_{ij} is the *Kronecker symbol*, equal to 1 when the two indices are equal and vanishing otherwise.

The whole apparatus can be condensed in a system of three coupled second-order PDEs for the three displacement components, namely,

$$C_{ijkl}\, u_{k,lj} + b_i = \rho \frac{\partial^2 u_i}{\partial t^2} \qquad i = 1, 2, 3. \tag{7.83}$$

7.2.5.2 Hyperbolicity

Introducing the wave front equation $\phi(x_1, x_2, x_3, t) = \psi(x_1, x_2, x_3) - t = 0$ and implementing Eq. (7.36), we obtain the *propagation condition*

$$\left(C_{ijkl} m_j m_l - \rho U^2 \delta_{ik} \right) a_k = 0, \tag{7.84}$$

in which m_i are the components of a unit vector normal to the wave front and a_i are the components of the *wave amplitude vector*. The tensor

$$Q_{ik} = C_{ijkl} m_j m_l \tag{7.85}$$

is called the *acoustic tensor in the direction* **m**.[3] Total hyperbolicity corresponds to the case in which, for every **m**, the tensor **Q** is (symmetric and) positive-definite with three distinct eigenvalues proportional to the speeds of propagation. The associated orthonormal eigenvectors are called the *acoustical axes* associated with the direction **m**. If one of the acoustical axes coincides with **m**, the wave is said to be *longitudinal*.

[3] For a fuller treatment of the general non-linear theory, see [6, 7].

If they are perpendicular, the wave is called *transversal*. These concepts make sense because $n = K - 1$ and the displacement vector can be regarded as an element of the physical space $\mathbb{R}^3$. The case of one repeated eigenvalue occurs in isotropic materials, where transversal waves propagate at identical speed.

Exercises

Exercise 7.1 Write the equations of an intermediate theory of beams for which the normality of cross sections is preserved but the rotary inertia is included. Find the speed of propagation of weak singularities for this theory.

Exercise 7.2 In the Timoshenko beam we found 4 distinct speeds of propagation and identified the first two as pertaining to bending waves and the last two as pertaining to shear waves. Justify this terminology by calculating the eigenvectors corresponding to each of the 4 speeds.

Exercise 7.3 Carry out all the steps necessary to obtain Eq. (7.16) and provide explicit expressions for $\hat{\mathbf{C}}$ and $\hat{\mathbf{d}}$. Compare with the results in [4], p. 48.

Exercise 7.4 Apply the procedure explained in Sect. 7.1.3 to the Timoshenko beam equations to obtain a system in canonical form. Consider the case of repeated eigenvalues and obtain the corresponding transport equations. Compare your results with those obtained in [5] and with those obtained in Box 7.3.

Exercise 7.5 It is not always possible to obtain a single higher order PDE equivalent to a given system of first-order PDEs. In the case of the Timoshenko beam, however, this can be done by differentiation and elimination. Provide two alternative formulations based, respectively, on a system of two second-order equations for the rotation θ and the deflection w, and on a single fourth-order equation for the deflection w. Obtain the characteristic speeds from each of these two alternative formulations.

Exercise 7.6 Prove that the speed of propagation of a surface moving on a material background, as described in Box 7.2, is given by

$$U_p = -\frac{Df/Dt}{\sqrt{\nabla f \cdot \nabla f}},$$

where D/Dt is the material derivative operator defined in Box 2.1.

Exercise 7.7 Show that in an isotropic elastic material (in infinitesimal elasticity) all transverse waves travel at the same speed. Find the speeds of longitudinal and transverse waves in terms of the Lamé constants. Determine restrictions on these constants so that waves can actually exist.

References

1. Cohen H, Epstein M (1982) Wave fronts in elastic membranes. J Elast 12(3):249–262
2. Gentle JE (2007) Matrix algebra. Springer, Berlin
3. Hutchinson JR (2001) Shear coefficients for Timoshenko beam theory. ASME J Appl Mech 68:87–92
4. John F (1982) Partial differential equations. Springer, Berlin
5. Leonard RW, Budiansky B (1954) On traveling waves in beams, NACA report 1173. http://naca.central.cranfield.ac.uk/reports/1954/naca-report-1173.pdf
6. Truesdell C, Noll W (1965) The non-linear field theories of mechanics. In: Flügge S (ed) Handbuch der Physik. Springer, Berlin
7. Varley E, Cumberbatch E (1965) Non-linear theory of wave-front propagation. J Inst Maths Applics 1:101–112

Part IV
Paradigmatic Equations

Chapter 8
The One-Dimensional Wave Equation

The archetypal hyperbolic equation is the wave equation in one spatial dimension. It governs phenomena such as the propagation of longitudinal waves in pipes and the free transverse vibrations of a taut string. Its relative simplicity lends itself to investigation in terms of exact solutions of initial and boundary-value problems. The main result presented in this chapter is the so-called d'Alembert solution, expressed within any convex domain as the superposition of two waves traveling in opposite directions with the same speed. Some further applications are explored.

8.1 The Vibrating String

We have already encountered the one-dimensional wave equation in Chap. 2 when dealing with longitudinal deformations of an elastic bar of constant cross section. There, we obtained the second-order PDE

$$u_{tt} = c^2\, u_{xx}, \tag{8.1}$$

where the constant c is given by

$$c = \sqrt{\frac{E}{\rho}}, \tag{8.2}$$

E being Young's modulus and ρ the density of the material.

For purposes of visualization, it is sometimes convenient to have in mind another phenomenon, also governed by Eq. (8.1), namely, the small transverse vibrations of a string under tension, just like the strings of a musical instrument. To derive the equation in this context, we will resort to an elementary technique. Denoting by A the cross-sectional area and by T the value of the tension, and assuming that the transverse deflections u are so small that the tension remains practically unaffected

M. Epstein, *Partial Differential Equations*, Mathematical Engineering,
DOI 10.1007/978-3-319-55212-5_8

Fig. 8.1 A vibrating string

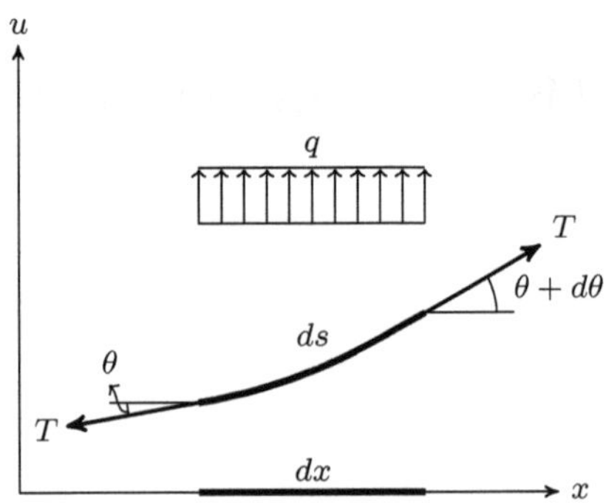

by the motion, we can draw a free-body diagram of an infinitesimal portion of the string of extent dx, as shown in Fig. 8.1.

Newton's second law applied to the vertical components of the forces yields

$$-T\ \sin\theta + T\ \sin(\theta + d\theta) + q\ dx = u_{tt}\ \rho\ A\ dx. \tag{8.3}$$

Since the angles are assumed to be very small, we can replace the sine and the tangent of the arc by the arc itself. Moreover, at any fixed instant of time

$$d\theta \approx d(\tan\theta) = du_x = u_{xx}dx. \tag{8.4}$$

Putting all these results together, we obtain

$$u_{tt} = c^2\ u_{xx} + q. \tag{8.5}$$

The constant c is now given by

$$c = \sqrt{\frac{T}{\rho\ A}}. \tag{8.6}$$

As before, this constant has the units of speed. In the absence of externally applied transverse forces q, we recover the one-dimensional wave equation (8.1). We refer to this case as the *free vibrations* of the string.

8.2 Hyperbolicity and Characteristics

The characteristic equation (6.18) is given, in the case of the wave equation and with the notation of the present chapter, by

$$\frac{dt}{dx} = \pm\frac{1}{c}. \tag{8.7}$$

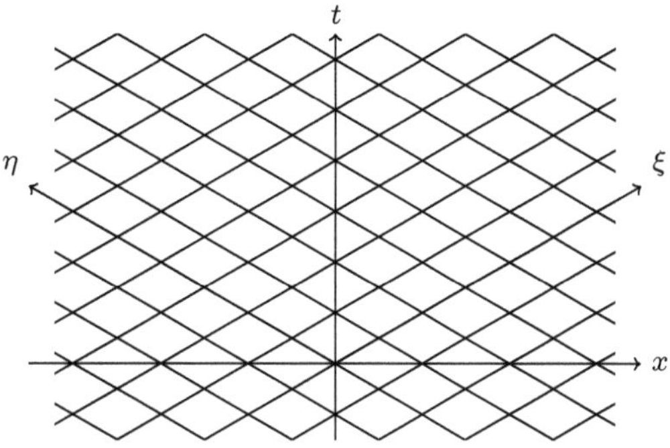

Fig. 8.2 Characteristics as coordinate *lines* ξ, η

In other words, the equation is hyperbolic. Since the equation is linear, the projected characteristics are, as expected, independent of the solution. Moreover, because the equation has constant coefficients, they are straight lines. We have seen that weak singularities can occur only across characteristics. It follows in this case that weak signals[1] propagate with the constant speed c. Introducing the new variables

$$\xi = \frac{x + c\,t}{2} \qquad \eta = \frac{x - c\,t}{2}, \tag{8.8}$$

the one-dimensional wave equation is reduced to the canonical (normal) form

$$u_{\xi\eta} = 0. \tag{8.9}$$

This change of variables is tantamount to using the characteristics as coordinate lines, as suggested by Fig. 8.2.

8.3 The d'Alembert Solution

Suppose that we are given a convex[2] domain in $\mathbb{R}^2$ and a C^2 function $u = u(\xi, \eta)$ satisfying the wave equation identically in that domain. Since, according to the wave equation (8.9), we must have that

$$u_{\xi\eta} = \frac{\partial}{\partial\eta}\left(\frac{\partial u(\xi, \eta)}{\partial\xi}\right) = 0, \tag{8.10}$$

[1] As discussed in Sect. 7.1.3, this applies to strong signals as well.

[2] A subset of $\mathbb{R}^n$ is convex if whenever two points belong to the set so does the entire line segment comprised by them. The requirement of convexity is sufficient for the validity of the statement. If the domain is not convex, we may have different functions in different parts of the domain, and still satisfy the differential equation.

we conclude that, in the domain of interest, there exists a function $F(\xi)$ such that

$$\frac{\partial u(\xi, \eta)}{\partial \xi} = F(\xi). \tag{8.11}$$

A further integration of this equation yields

$$u(\xi, \eta) = f(\xi) + g(\eta), \tag{8.12}$$

where f and g are arbitrary C^2 functions of one real variable.

It is not difficult to verify that, for a given solution $u = u(\xi, \eta)$, the functions f and g are unique to within a constant (that, if added to f, must be subtracted from g). Returning to the original variables x, t, we can write the general solution of Eq. (8.1) as

$$u(x, t) = f(x + c\,t) + g(x - c\,t). \tag{8.13}$$

This is the so-called *d'Alembert solution* of the one-dimensional wave equation. In some sense, it states that we can interpret the wave equation as the second-order counterpart of the advection equation discussed in Chap. 3. Indeed, it establishes that the general solution of the one-dimensional wave equation is a superposition of two waves propagating with equal speed c in opposite directions while keeping their shapes unchanged in time.

8.4 The Infinite String

The d'Alembert representation of the general solution applies to any convex domain. Nevertheless, the solution of any real-life problem requires the specification of boundary (fixed ends of a finite string) and initial (position and velocity at $t = 0$) conditions. In particular, the boundary conditions limit the direct use of the d'Alembert representation over the whole domain of interest. For this reason, the best direct application of the d'Alembert solution is to the case of a spatially unbounded domain (i.e., an infinite string). Assume that the initial conditions are given by some functions $u_0(x)$, $v_0(x)$, namely,

$$u(x, 0) = u_0(x) \tag{8.14}$$

and

$$u_t(x, 0) = v_0(x). \tag{8.15}$$

In other words, the functions $u_0(x)$, $v_0(x)$ represent, respectively, the known initial shape and velocity profile of the string. Using the d'Alembert representation (8.13), we immediately obtain

$$u_0(x) = f(x) + g(x) \tag{8.16}$$

and

$$v_0(x) = c\, f'(x) - c\, g'(x), \tag{8.17}$$

where primes are used to denote derivatives of a function of a single variable. Differentiating equation (8.16) and combining the result with Eq. (8.17), we can read off

$$f'(x) = \frac{1}{2} u_0'(x) + \frac{1}{2c} v_0(x), \tag{8.18}$$

$$g'(x) = \frac{1}{2} u_0'(x) - \frac{1}{2c} v_0(x). \tag{8.19}$$

On integrating, we obtain

$$f(x) = \frac{1}{2} u_0(x) + \frac{1}{2c} \int_0^x v_0(z)\, dz + C, \tag{8.20}$$

$$g(x) = \frac{1}{2} u_0(x) - \frac{1}{2c} \int_0^x v_0(z)\, dz - C. \tag{8.21}$$

Notice that the lower limit of integration is immaterial, since it would only affect the value of the constant of integration C. The reason for having the same integration constant (with opposite signs) in both expressions stems from the enforcement of Eq. (8.16). Notice also that the dependence on x is enforced via the upper limit, according to the fundamental theorem of calculus, while z is a dummy variable of integration. For the functions f and g to be of class C^2 it is sufficient that u_0 be C^2 and that v_0 be C^1.

Now that we are in possession of the two component functions of the d'Alembert representation, we are in a position of stating the final result, namely,

$$\begin{aligned} u(x,t) &= f(x + c\,t) + g(x - c\,t) \\ &= \frac{1}{2} u_0(x + c\,t) + \frac{1}{2c} \int_0^{x+c\,t} v_0(z)\, dz + \frac{1}{2} u_0(x - c\,t) - \frac{1}{2c} \int_0^{x-c\,t} v_0(z)\, dz. \end{aligned} \tag{8.22}$$

or, more compactly,

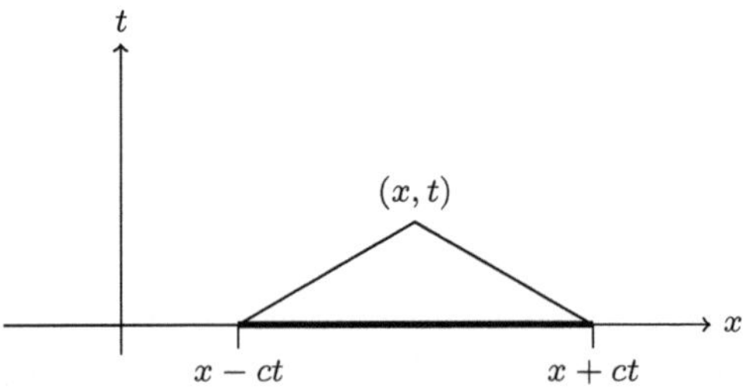

Fig. 8.3 Domain of dependence

$$u(x,t) = \frac{1}{2}u_0(x + c\,t) + \frac{1}{2}u_0(x - c\,t) + \frac{1}{2c}\int\limits_{x-c\,t}^{x+c\,t} v_0(z)\,dz. \qquad (8.23)$$

From the point of view of the general theory of hyperbolic equations, it is important to notice that the value of the solution at a given point (x, t) is completely determined by the values of the initial conditions in the finite closed interval $[x - c\,t, x + c\,t]$. This interval, called the *domain of dependence of the point* (x, t), is determined by drawing the two characteristics through this point backwards in time until they intersect the x axis, as shown in Fig. 8.3.[3] From the physical point of view, any initial datum or signal outside the domain of dependence of a point has no influence whatsoever on the response of the system at that point of space and time. In other words, signals propagate at a definite finite speed and cannot, therefore, be felt at a given position before a finite time of travel. This remark has implications on any numerical scheme of solution (such as the method of finite differences) that one may attempt to use in practice.

The natural counterpart of the notion of domain of dependence is that of *range of influence*. Given a closed interval $[a, b]$ (or, in particular, a point) in the x axis (or anywhere else, for that matter), its range of influence consists of the collection of points (in the future) whose domain of dependence intersects that interval (or point). It is obtained by drawing the outbound characteristics from the extreme points of the interval and considering the wedge shaped zone comprised between them, as shown in Fig. 8.4. A point outside the range of influence will be not affected at all by the initial data in the interval $[a, b]$.

[3]Further elaborations of these ideas can be found in most books on PDEs. For clarity and conciseness, we again recommend [3, 4].

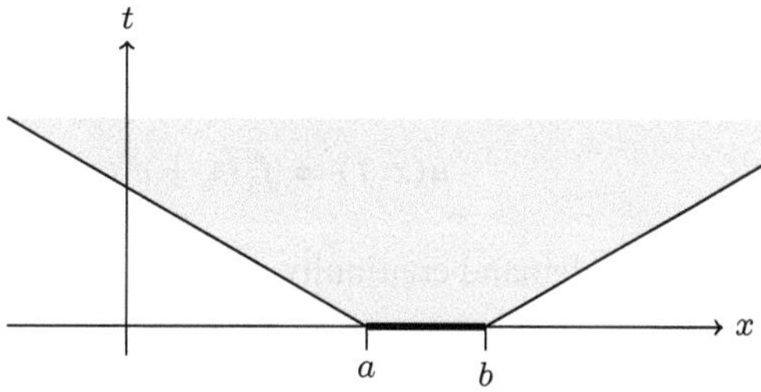

Fig. 8.4 Range of influence

8.5 The Semi-infinite String

8.5.1 D'Alembert Solution

The d'Alembert solution can also be applied to the problem of the dynamics of a semi-infinite string fixed or otherwise supported at one end. Let the undeformed string occupy the region $0 \leq x < \infty$ and let the initial conditions be given as in the previous problem by

$$u(x, 0) = u_0(x) \qquad 0 \leq x < \infty \tag{8.24}$$

and

$$u_t(x, 0) = v_0(x) \qquad 0 \leq x < \infty. \tag{8.25}$$

At the end $x = 0$ we must now impose a *boundary condition*. Consider the case in which this end is fixed, namely,

$$u(0, t) = 0 \qquad 0 \leq t < \infty. \tag{8.26}$$

Note that at time $t = 0$ the boundary condition may happen to be inconsistent with the initial conditions at the left end of the string. We will for now explicitly assume that this is not the case. In other words, we assume that

$$u_0(0) = 0 \qquad v_0(0) = 0. \tag{8.27}$$

According to our discussion on domain of dependence and region of influence, it should be clear that the solution in the region under the right-pointing characteristic through the origin coincides with the solution for an infinite string with the same initial conditions, viz.,

$$u(x, t) = \frac{1}{2}u_0(x + c\,t) + \frac{1}{2}u_0(x - c\,t) + \frac{1}{2c}\int\limits_{x-c\,t}^{x+c\,t} v_0(z)\,dz \qquad 0 \leq ct \leq x. \tag{8.28}$$

The solution in the remaining part of the domain (namely, $0 \leq x \leq ct$) must also be amenable to a d'Alembert decomposition of the form

$$u(x, t) = f_1(x + c\,t) + g_1(x - c\,t). \tag{8.29}$$

At $x = c\,t$ we demand continuity, that is,

$$f_1(2c\,t) + g_1(0) = f(2c\,t) + g(0). \tag{8.30}$$

Since we have a constant of integration at our disposal, we may set $g_1(0) = g(0)$ and obtain

$$f_1(x) = f(x). \tag{8.31}$$

The physical meaning of this result is that the backward moving wave is unaffected by the presence of the support, which should not be surprising. To obtain the forward moving wave, we impose the boundary condition (8.26) and obtain

$$0 = u(0, t) = f_1(c\,t) + g_1(-c\,t) \qquad \forall t > 0. \tag{8.32}$$

In other words,

$$g_1(x - c\,t) = -f_1(-(x - c\,t)) \qquad 0 \leq x < c\,t. \tag{8.33}$$

From the physical point of view this result means that the forward wave arriving at a point with coordinates x_0, t_0 situated in the upper domain is a reflected (and inverted) version of the backward wave issuing at the initial time $t = 0$ from the point of the string with coordinate $ct_0 - x_0$. The time at which the reflection occurs is $t_0 - x/c$, as shown in Fig. 8.5.

Combining all the above results, we can obtain an explicit formula for the solution in the upper domain as

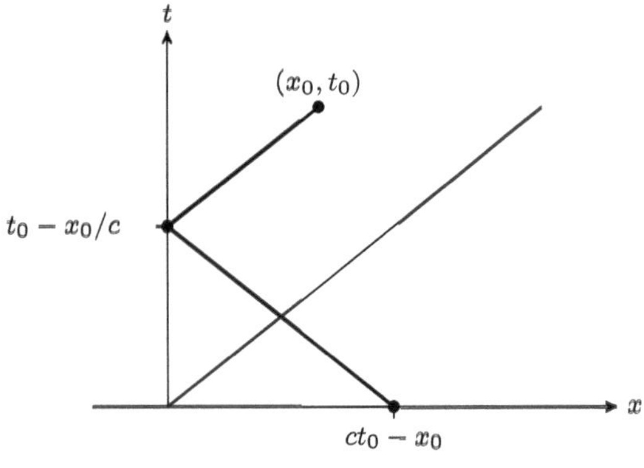

Fig. 8.5 Reflection of backward waves in the semi-infinite string

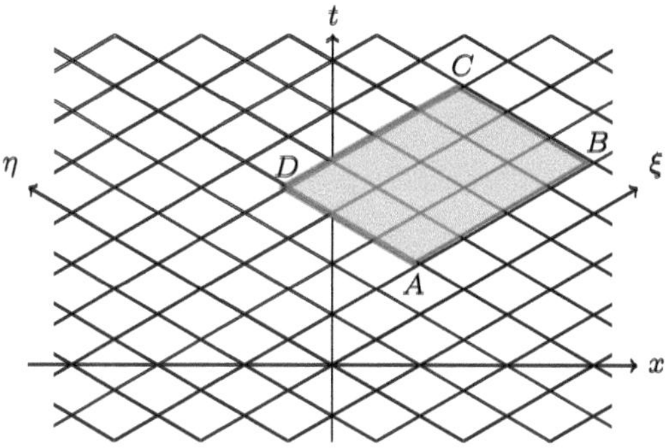

Fig. 8.6 A characteristic parallelogram

$$u(x,t) = \frac{1}{2}u_0(x + c\,t) - \frac{1}{2}u_0(-(x - c\,t)) + \frac{1}{2c}\int\limits_{-(x-c\,t)}^{x+c\,t} v_0(z)\,dz \qquad 0 \le x \le c\,t. \tag{8.34}$$

8.5.2 Interpretation in Terms of Characteristics

The representation used in Fig. 8.5 suggests that the analytic expressions obtained by means of d'Alembert's solution can be also obtained geometrically by constructions based on the characteristic lines alone. To this end, consider a parallelogram-shaped domain enclosed by characteristics, such as the shaded domain shown in Fig. 8.6.

Denoting the corners of the parallelogram by A, B, C, D, as shown, it is not difficult to conclude that any function $u = u(x,t)$ of the d'Alembert form (8.13) satisfies the condition

$$u_A + u_C = u_B + u_D, \tag{8.35}$$

with an obvious notation. This result follows from Eq. (8.12) on observing that

$$\xi_A = \xi_D \qquad \xi_B = \xi_C \qquad \eta_A = \eta_B \qquad \eta_C = \eta_D. \tag{8.36}$$

Every C^2 solution of the wave equation, therefore, satisfies the simple algebraic identity (8.35). Conversely, every function that satisfies Eq. (8.35) for every characteristic parallelogram within a given domain also satisfies the wave equation within the domain.[4] If we have a function that is not of class C^2 and yet satisfies Eq. (8.35),

[4]See [3], p. 42.

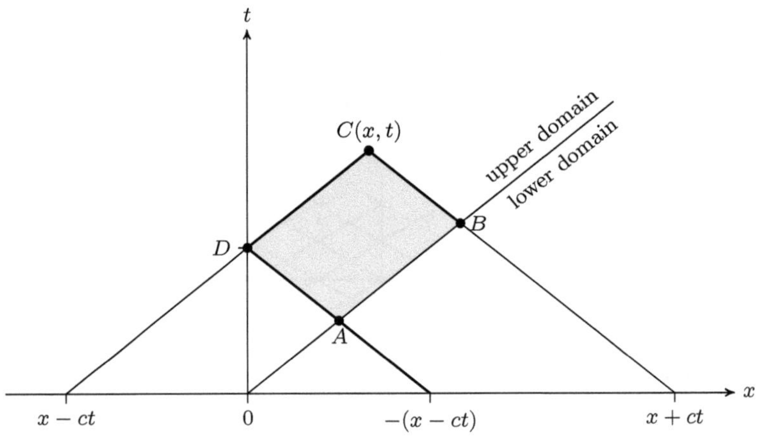

Fig. 8.7 Geometrical derivation of the solution

we may say that it satisfies the wave equation in a *weak sense* or, equivalently, that it is a *generalized* or *weak solution* of the one-dimensional wave equation.

Let us apply Eq. (8.35) to the solution of the semi-infinite string problem with a fixed end. Consider a point C with coordinates (x, t) in the upper domain of the problem, as represented in Fig. 8.7. We complete a characteristic parallelogram by drawing the two characteristics from this point back to the lines $x = 0$ and $x = c\,t$, thus obtaining the points D and B, respectively. The remaining point, A, is obtained as the intersection of the characteristic line issuing from D and the line $x = c\,t$, as shown in the figure.

By the boundary condition (8.26), we have

$$u_D = 0. \tag{8.37}$$

Therefore, by Eq. (8.35), we must have

$$u(x, t) = u_C = u_B - u_A. \tag{8.38}$$

If we demand continuity of the solution, however, the values of u_A and u_B can be obtained by integrating over their respective domains of dependence according to Eq. (8.23). Thus,

$$u_A = \frac{1}{2}\,u_0(-(x - c\,t)) + \frac{1}{2}\,u_0(0) + \frac{1}{2c}\int_0^{-(x-c\,t)} v_0(z)\,dz \tag{8.39}$$

$$u_B = \frac{1}{2}\, u_0(x + c\,t) + \frac{1}{2}\, u_0(0) + \frac{1}{2c} \int\limits_0^{x+c\,t} v_0(z)\, dz. \tag{8.40}$$

Combining Eqs. (8.38), (8.39) and (8.40) we recover the solution (8.34).

8.5.3 Extension of Initial Data

There exists yet another way to obtain the solution for a semi-infinite (and, eventually, for a finite) string. It consists of extending the initial data to the whole line in such a way that the boundary conditions are satisfied automatically by the corresponding solution of the infinite string. This procedure needs to be used with care and on a case-by-case basis. Consider the case discussed in the previous section, namely, the one with initial and boundary conditions described by Eqs. (8.24)–(8.27). We now extend the initial conditions as odd functions over the whole domain. Such an extension is shown pictorially in Fig. 8.8, where the initial function, $u_0(x)$ or $v_0(x)$, given in the original domain $0 \le x < \infty$, has been augmented with a horizontally and vertically flipped copy so as to obtain an odd function over the whole domain.

The extended functions, denoted here with a bar, coincide with the given data in the original domain and, moreover, enjoy the property

$$\bar{u}_0(x) = -\bar{u}_0(-x) \qquad \bar{v}_0(x) = -\bar{v}_0(-x) \qquad -\infty < x < \infty. \tag{8.41}$$

In other words, they are odd functions of their argument.

Let $\bar{u}(x, t)$ denote the solution to the problem of the infinite string under the initial conditions $\bar{u}_0(x)$, $\bar{v}_0(x)$. In accordance with Eq. (8.23), this solution is given by

$$\bar{u}(x, t) = \frac{1}{2}\, \bar{u}_0(x + c\,t) + \frac{1}{2}\, \bar{u}_0(x - c\,t) + \frac{1}{2c} \int\limits_{x-c\,t}^{x+c\,t} \bar{v}_0(z)\, dz. \tag{8.42}$$

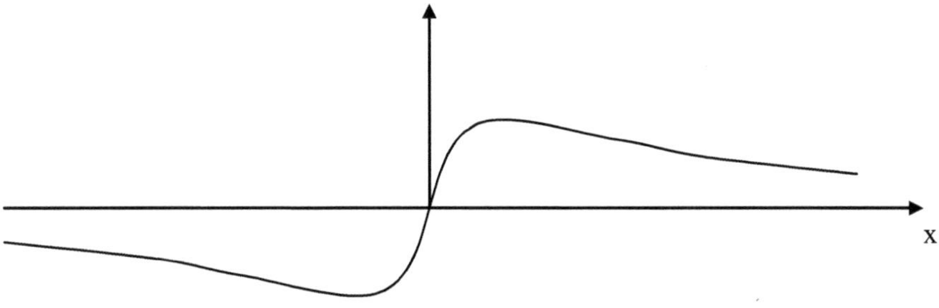

Fig. 8.8 Extension of data for the semi-infinite string

Let us evaluate this solution over the positive time axis. The result is

$$\bar{u}_0(0,t) = \frac{1}{2}\,\bar{u}_0(c\,t) + \frac{1}{2}\,\bar{u}_0(-c\,t) + \frac{1}{2c}\int\limits_{-c\,t}^{c\,t} \bar{v}_0(z)\,dz. \tag{8.43}$$

By virtue of the deliberately imposed condition (8.41), we immediately obtain

$$\bar{u}(0,t) = 0. \tag{8.44}$$

We conclude that the extended solution automatically satisfies the desired boundary condition of the semi-infinite string. It follows, therefore, that restricting the extended solution to the original domain provides the solution[5] to the original problem. From the physical point of view, we may say that the forward and backward waves of the extended solution interfere with each other destructively, in a way that has been precisely calibrated to produce a zero value for all times at $x = 0$.

It is interesting to note that the solution obtained will in general not be of class C^2 since, even if the original initial conditions and the extended initial conditions are C^2 and even if they vanish at the origin, the extended initial conditions are only guaranteed to be C^1 at the origin, unless the extra conditions

$$u_0''(0) = v_0''(0) = 0 \tag{8.45}$$

happen to be satisfied. If these conditions are not satisfied, the corresponding weak discontinuities will propagate along the characteristic emerging from the origin. This feature is not due to the method of solution (since the solution is ultimately unique).

Remark 8.1 Although the method of extension of the initial conditions is a legitimate alternative to the method of characteristics, it should be clear that the latter is more general than the former. Indeed, the method of extension of the initial data is applicable only when the supported end is actually fixed, whereas the method of characteristics is still viable when the support is subjected to a given motion.

8.6 The Finite String

8.6.1 Solution

The method of characteristics can be used in principle to solve for the vibrations of a finite string of length L with appropriate boundary conditions. All one has to do is to divide the domain of interest (which consists of a semi-infinite vertical strip

[5]Clearly, we are tacitly invoking an argument of uniqueness, which we have not pursued.

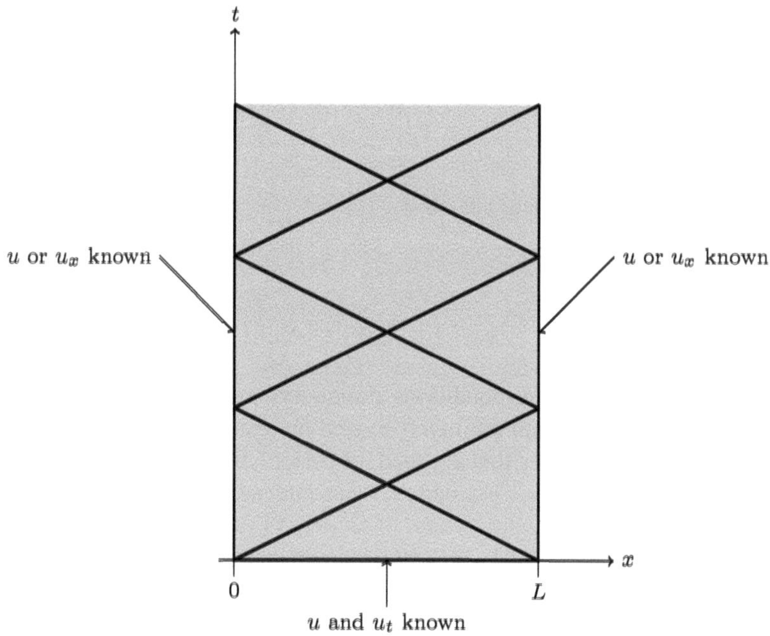

Fig. 8.9 Characteristic solution for the finite string

in space-time) into regions, as shown in Fig. 8.9. The lower triangular region at the base of the strip is solved by the general formula (8.23). Assuming that the solution is continuous over the whole strip, we can then use the values at the boundary of this triangle, together with the known values at the vertical boundaries of the strip, to use the parallelogram formula (8.35) for the next two triangular regions. The procedure is carried out in a similar manner for increasingly higher (triangular or parallelogram) regions. It is not difficult to write a computer code to handle this recursive algorithm (as suggested in Exercise 8.10).

For the particular case of fixed ends, an equivalent procedure is obtained by exploiting the extension idea, namely the 'trick' already used for the semi-infinite string to reduce the problem to that of an infinite string. Consider the problem of a finite string of length L whose ends are fixed, namely,

$$u(0,t) = 0 \qquad u(L,t) = 0 \qquad \forall t \geq 0, \tag{8.46}$$

with given initial conditions

$$u(x,0) = u_0(x) \qquad u_t(0) = v_0(x) \qquad 0 \leq x \leq L. \tag{8.47}$$

To avoid any possible strong discontinuities, we confine our attention to the case in which the initial and boundary conditions are consistent with each other in the sense

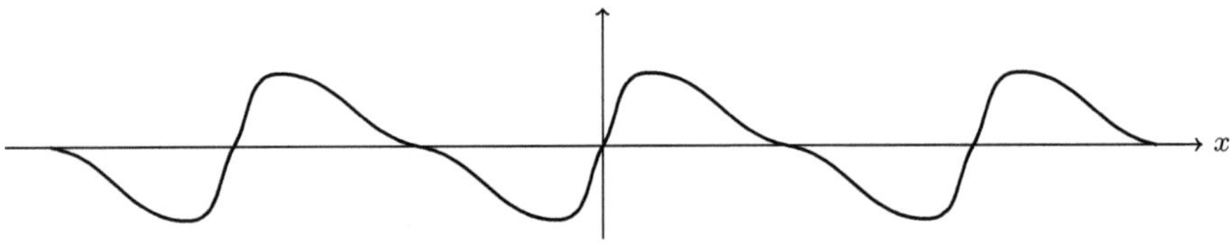

Fig. 8.10 Extension of data for the finite string

that

$$u_0(0) = u_0(L) = v_0(0) = v_0(L) = 0. \tag{8.48}$$

We now extend the initial conditions (uniquely) to odd functions with period $2L$. Although this idea is both intuitively appealing and, eventually, justified by the results, it is worthwhile noting that the need for periodicity can actually be derived by means of a rational argument.[6] The odd periodic extension of a function is illustrated in Fig. 8.10.

In addition to conditions (8.41), the extended initial data satisfy now the periodicity conditions

$$\bar{u}_0(x) = \bar{u}_0(x + 2L) \qquad \bar{v}_0(x) = \bar{v}_0(x + 2L) \qquad -\infty < x < \infty. \tag{8.49}$$

By the same procedure as in the case of the infinite string, invoking just the odd character of the extension, we can be assured that the fixity condition is satisfied at the left end of the string. The periodicity takes care of the right end. Indeed

$$\bar{u}(L, t) = \frac{1}{2}\, \bar{u}_0(L + c\, t) + \frac{1}{2}\, \bar{u}_0(L - c\, t) + \frac{1}{2c} \int_{L-c\,t}^{L+c\,t} \bar{v}_0(z)\, dz. \tag{8.50}$$

But, for odd periodic functions, we have

$$\bar{u}_0(L + x) = -\bar{u}_0(L - x) \qquad \bar{v}_0(L + x) = -\bar{v}_0(L - x). \tag{8.51}$$

Combining this fact with Eq. (8.50), we immediately obtain

$$\bar{u}(L, t) = 0. \tag{8.52}$$

Again, from the physical point of view, we can interpret this result as the outcome of a carefully calibrated mutually destructive interference of the backward and forward traveling waves.

[6]See [6], p. 50, a text written by one of the great Russian mathematicians of the 20th century.

8.6.2 Uniqueness and Stability

We have found solutions to the wave equation with given initial and boundary conditions by means of different procedures. Is the solution *unique*? Is it *stable* in the sense that small changes in the initial and/or boundary conditions result in small changes in the solution? These questions are of great importance for PDEs in general and must be studied in detail for each case. A problem in PDEs is said to be *well posed* if it can be shown that a solution exists, that it is unique and that it depends continuously (in some sense) on the initial and boundary data. For the one-dimensional wave equation on a finite spatial domain the answer to the question of uniqueness can be found, somewhat unexpectedly, in the underlying physics of the problem by invoking the principle of conservation of energy in non-dissipative systems, as explained in Box 8.1.

We need to verify that, from the mathematical viewpoint, the total energy $W = K + U$ is indeed a conserved quantity. Invoking the results of Box 8.1, namely,

$$W = \frac{1}{2}\int_0^L \left(u_t^2 + c^2u_x^2\right)\rho A dx, \tag{8.53}$$

we calculate the rate of change of the energy on any C^2 solution $u = u(x, t)$ of the wave equation as

$$\begin{aligned} \frac{dW}{dt} &= \frac{1}{2}\frac{d}{dt}\int_0^L \left(u_t^2 + c^2u_x^2\right)\rho A dx \\ &= \int_0^L \left(u_t u_{tt} + c^2 u_x u_{xt}\right)\rho A dx \\ &= \int_0^L \left(u_t u_{tt}\right)\rho A dx + \left[c^2 u_x u_t \rho A\right]_{x=0}^{x=L} - \int_0^L u_t c^2 u_{xx}\rho A dx \\ &= \int_0^L u_t\left(u_{tt} - c^2 u_{xx}\right)\rho A dx + \left[c^2 u_x u_t \rho A\right]_{x=0}^{x=L} \\ &= \left[c^2 u_x u_t \rho A\right]_{x=0}^{x=L}, \end{aligned} \tag{8.54}$$

where we have integrated by parts and assumed, for simplicity, a constant density ρ and a constant cross section A. For the case of fixed ends, since $u_t = 0$ at both ends, we obtain the desired result, namely,

Box 8.1 Energy

The total energy can be split into *kinetic* and *potential* components. When the wave equation is derived in the context of longitudinal waves in elastic bars (as done in Sect. 2.4.3), the kinetic energy due to the longitudinal translation of the cross sections is given by

$$K = \int_0^L \frac{1}{2}\rho u_t^2 A dx,$$

where we are confining our attention to the case of a finite bar of length L. The total potential energy is stored as elastic energy, just as in the case of a linear spring, given by

$$U = \int_0^L \frac{1}{2} E u_x^2 A dx = \int_0^L \frac{1}{2}\rho c^2 u_x^2 A dx.$$

For the application to the transverse vibrations of a taut string, the kinetic energy K is given by the same expression as its counterpart for longitudinal waves, except that the variable u must be interpreted as a transverse, rather than longitudinal, displacement. The potential energy U for the vibrating string, on the other hand, must be examined more carefully. Indeed, we have assumed that the tension T remains constant during the process of deformation.[a] It does, however, perform work by virtue of the extension of the string. If ds represents the length of a deformed element, whose original length was dx, as indicated in Fig. 8.1, we can write

$$ds = \sqrt{1 + u_x^2}dx = \left(1 + \frac{1}{2}u_x^2 + \ldots\right) dx.$$

The infinitesimal work performed by the tension T is, therefore,

$$T(ds - dx) \approx \frac{1}{2} T u_x^2 = \frac{1}{2}\rho A c^2 u_x^2.$$

Since this work is completely recovered upon return to the straight configuration, it can legitimately be called a potential energy and we obtain the same expression as in the case of the longitudinal waves in a bar.

[a]The energy is, of course, of elastic origin. This is a case of small deformations imposed upon pre-existent larger ones. As an example, consider a linear spring with elongation e and internal force $F = ke$, whose elastic energy is $W = 0.5ke^2$. Its increment due to a small de superimposed on a background e_0 is $dW = ke_0 de = F de$.

$$\frac{dW}{dt} = 0. \tag{8.55}$$

The same result is, of course, obtained if an end is free to move but subjected to no external force, so that $u_x = 0$.

Let $u_1 = u_1(x, t)$ and $u_2 = u_2(x, t)$ be C^2 solutions of the wave equation with the same boundary and initial conditions. Since the wave equation is linear, the difference $u = u_1 - u_2$ satisfies the wave equation with homogeneous (that is, vanishing) boundary and initial conditions. Thus, the total energy at time $t = 0$ vanishes. Since, as we have just demonstrated, the total energy is conserved, we have

$$\frac{1}{2}\int_0^L \left(u_t^2 + c^2 u_x^2\right) \rho A dx = 0 \tag{8.56}$$

for all times. In view that the integrand is non-negative, this result is only possible if it vanishes identically. Hence, $u = u_1 - u_2$ is constant and, therefore, zero. This concludes the proof of uniqueness.

The issue of continuous dependence on initial data can be settled, for instance, by using the technique of periodic extension and explicitly determining a norm of the difference between the solutions corresponding to two sets of initial data. This issue is clearly of great importance for numerical methods of solution of PDEs.

8.6.3 *Time Periodicity*

The free vibrations of a finite string are necessarily periodic in time. This fact can be established in various ways, one of which we will pursue presently. We know that the (unique) solution for the free vibrations of a simply supported string of length L can be obtained by the technique of extending the initial conditions periodically over $\mathbb{R}$. The extension of the initial conditions u_0 and v_0 yields, respectively, two *odd* functions $\bar{u}_0(x)$ and $\bar{v}_0(x)$ with a period of $2L$ and the (d'Alembert) solution of the problem is obtained as

$$u(x,t) = \frac{1}{2}\bar{u}_0(x + c\,t) + \frac{1}{2}\bar{u}_0(x - c\,t) + \frac{1}{2c}\int_{x-c\,t}^{x+c\,t} \bar{v}_0(z)\,dz. \tag{8.57}$$

For each value of x, this function $u(x,t)$ turns out to be periodic in time, with a period of $2L/c$, namely,

$$u(x, t + 2L/c) = u(x,t), \tag{8.58}$$

as can be verified by direct substitution noting that, due to the assumed odd character of the extension, the integral of the initial velocity over a whole period must vanish.

Notice that, although the displacement pattern is recovered exactly at regular intervals equal to the period, its shape, in general, varies with time, neither does the string become instantaneously un-deformed at any intermediate instant. We will later show that there are special solutions that do preserve their shapes and vary only in amplitude as time goes on. These special solutions can be interpreted physically as *standing waves*. They provide a different avenue of approach to the solution of vibration problems in Engineering and Physics.

8.7 Moving Boundaries and Growth

A violin player can change continuously the length of the string by sliding a finger over the fingerboard in a smooth motion known as *glissando*. This change of length involves an increase in the material content of the string enclosed between the two supports. Similarly, biological growth may arise as the result of additional material being deposited at the boundaries of an organ. These two processes involve time scales much larger than, say, the period of free vibrations of the original body. Nevertheless, they are interesting pictures to bear in mind when dealing with problems of moving boundaries. In the case of the violin player, the effect of the moving boundary can be perceived clearly by the ear as a variation of the fundamental pitch of the sound. Remarkably, the method of characteristics can be applied without any essential modification to problems involving moving boundaries.

Consider the idealized 'violinist problem' consisting of solving the wave equation in the domain $\mathcal{D}$ (shaded in Fig. 8.11) with initial conditions

$$u(x,0) = u_0(x) \qquad u_t(0) = 0 \qquad 0 \le x \le L, \tag{8.59}$$

and with boundary conditions

$$u(0,t) = 0 \qquad u(L+\alpha t, t) = 0 \qquad \forall t \ge 0. \tag{8.60}$$

In this equation, α is assumed to be a positive constant, so that the vibrating length of the string increases linearly with time. The constant α represents the speed of translation of the right support. On intuitive grounds, to be verified by the mathematics, this speed should be smaller than the speed c for the problem to have a solution different from the restriction to $\mathcal{D}$ of the solution of the semi-infinite string.

The algorithm to compute the value of the solution at any point $P \in \mathcal{D}$ can be conceived as follows: (a) Starting at P draw the downward-right characteristic until it intersects the boundary of $\mathcal{B}$; (b) if this intersection falls at a point A within the base segment $0 \le x \le L$, stop; otherwise (c) reflect the line by changing the sign

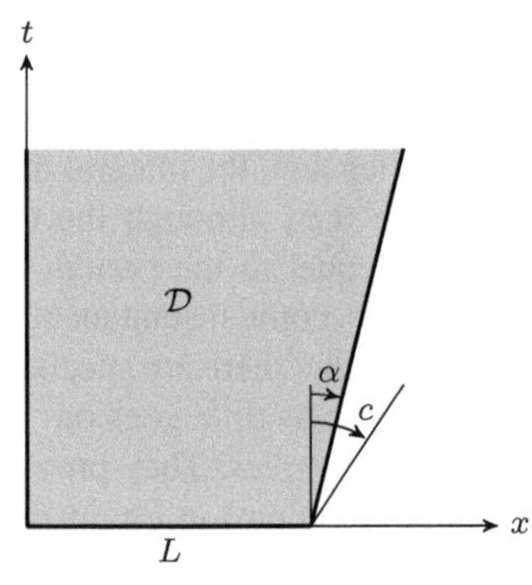

Fig. 8.11 Domain of definition for the moving boundary problem

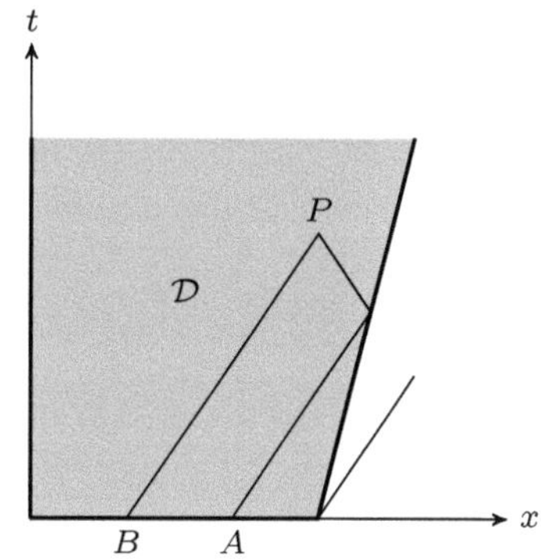

Fig. 8.12 Solution algorithm

of the slope and repeat as many times as needed to arrive at a point A in the base segment; (d) keep track of the number of bounces n_R; (e) repeat the whole zigzag procedure but starting with the downward-left characteristic to obtain a point B in the base after a number n_L of bounces (if any); the value of the displacement at P is given by

$$u_P = \frac{1}{2}\left((-1)^{n_R}\, u_0(A) + (-1)^{n_L}\, u_0(B)\right). \tag{8.61}$$

In the example of Fig. 8.12, we have $n_R = 1$ and $n_L = 0$.

8.8 Controlling the Slinky?

The Slinky is one of the most celebrated toys of the twentieth century. Essentially a helical spring, its versatility is due in part to the interaction of its elasticity with gravity and with lateral as well as axial deformations. To simplify matters, though, let us consider it as a linear elastic bar with density ρ, modulus of elasticity E and un-stretched length L undergoing axial displacements only in the small deformation regime under no external forces.[7] Under these conditions, it abides by the wave equation. The question we pose is the following: Holding one end while the other end is free and starting from a rest configuration, is it possible to obtain a prescribed displacement history $f(t)$ of the free end by imposing a displacement history $g(t)$ of the held end? From the mathematical standpoint, this is a problem in the theory of boundary control of PDEs. Here, however, we will consider it as an independent challenge for its own sake.

The key to solve this problem is that, for the case of the wave equation, the roles of the space (x) and time (t) variables can be exchanged. What this means is that the following is a perfectly well-defined Cauchy problem: Solve the wave equation

[7]This simplified model is not realistic for the actual Slinky for many reasons. We are only using it as a motivation for a well defined problem in linear one-dimensional elasticity.

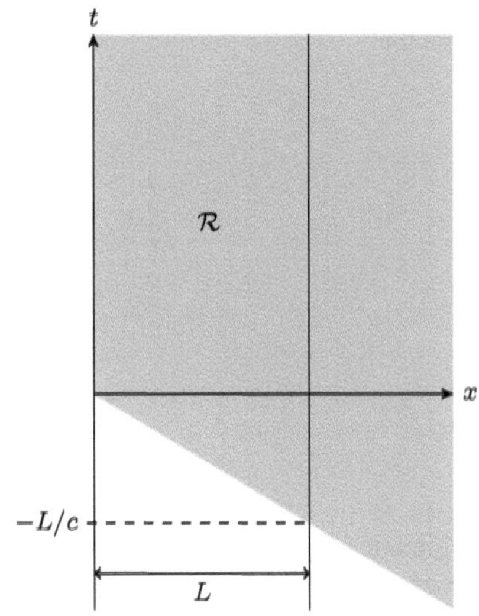

Fig. 8.13 Boundary control and the Slinky problem

$$u_{tt} - c^2 u_{xx} = 0 \qquad \text{for } x > 0, \tag{8.62}$$

with the 'initial' data

$$u(0, t) = U_0(t) \qquad u_x(0, t) = \varepsilon_0(t), \tag{8.63}$$

where $U_0(t)$ and $\varepsilon_0(t)$ are defined over the whole real line. Let us identify the free end with $x = 0$ and the held end with $x = L$. Let the desired displacement history of the free end be of the form

$$U_0(t) = \begin{cases} 0 & \text{for } t < 0 \\ f(t) & \text{for } t \geq 0 \end{cases} \tag{8.64}$$

and let $\varepsilon_0(t) = 0$, which corresponds to a zero strain (and stress) at an unsupported end, as desired.

The shaded area in Fig. 8.13 indicates the range of influence $\mathcal{R}$ of the non-vanishing Cauchy data. In particular, the line $x = L$ is affected only for times $t \geq -L/c$. As expected on physical grounds, therefore, the non-vanishing history $g(t)$ of the displacement to be applied at the held end must start earlier than the desired displacement at the free end. The explicit form of this history is, according to the d'Alembert solution implemented in the exchanged time-space domain,

$$g(t) = u(L, t) = \frac{1}{2} U_0 \left(t - \frac{L}{c} \right) + \frac{1}{2} U_0 \left(t + \frac{L}{c} \right), \tag{8.65}$$

which indeed vanishes identically for $t < L/c$. By uniqueness, we conclude that if we return to the original space-time picture and at time $t = -L/c$ (or any earlier time) we prescribe zero initial conditions of both displacement and velocity and as boundary conditions we specify at $x = 0$ an identically zero strain and at $x = L$ a displacement equal to $g(t)$, we should recover the same $u(x, t)$ for the finite beam as the one provided by the solution of the previous Cauchy problem. In particular, along the line $x = 0$, we should recover the desired displacement $f(t)$. The problem has been thus completely solved. For more realistic applications, it is not difficult to incorporate the effect of gravity.

A similar procedure can be used to demonstrate the boundary control of a Timoshenko beam. A good way to imagine this problem is to think of holding a fishing rod at one end and, by applying displacements and/or rotations at the held end, to try to achieve a specified displacement at the free end. By reversing the roles of space and time, it is possible to show [1] that, if both the displacement and the rotation at the held end are amenable to independent prescription, then the problem has a unique solution, just like the case of the slinky. When the displacement at the held end vanishes and only the rotation is amenable to prescription, the problem can also be solved, but requires the solution of a recursive functional equation.

8.9 Source Terms and Duhamel's Principle

Source terms give rise to the *inhomogeneous wave equation* in the form

$$u_{tt} - c^2\, u_{xx} = f(x, t), \tag{8.66}$$

that we have already encountered in deriving the equation for the vibrating string (8.5) subjected to an external force. It is enough to show how to solve this inhomogeneous problem with vanishing initial conditions and (for the finite case) vanishing boundary conditions, since any other conditions can be restored by superposition with the solution of the source-free case with any given initial and boundary conditions.

There are various ways to deal with the inhomogeneous wave equation.[8] The appeal of *Duhamel's principle* is that it can be motivated on physical, as well as mathematical, grounds. Mathematically, this method can be regarded as a generalization to linear PDEs of the method of *variation of constants* used in solving inhomogeneous linear ODEs. For a more intuitively physical motivation, see Box 8.2.

Assume that the solution of the homogeneous problem is available for arbitrary initial conditions. Moreover, the initial conditions could be specified, rather than just at $t = 0$, at any reference time $\tau > 0$ with the corresponding solution valid for $t \geq \tau$ and vanishing for $t < \tau$. In particular, we denote by $\bar{u}(x, t; \tau)$ the solution to the homogeneous problem with the initial conditions

[8] See, for example, [6] p. 51, [2] p. 103, [5] p. 221.

Box 8.2 The intuition behind Duhamel's principle
Duhamel's principle expresses the solution of an inhomogeneous PDE out of a clever continuous superposition of solutions of homogeneous problems, for which the solution is assumed to be known (either exactly or approximately). How is it possible to exchange a problem containing sources with a collection of source-free problems? A non-rigorous way to motivate Duhamel's principle consists of conceiving the effect of the right-hand side sources as the sum of the effects of sources with the same spatial distribution but which are constant in time over each one of a large number of small time intervals of extent $h = \Delta t$. In other words, we view the graph of the right-hand side of Eq. (8.66), representing the forcing function $f(x, t)$, sliced, as a slab of cheese divided into thin vertical slices, at regular time intervals of thickness $h = \Delta t$, as shown in the figure.

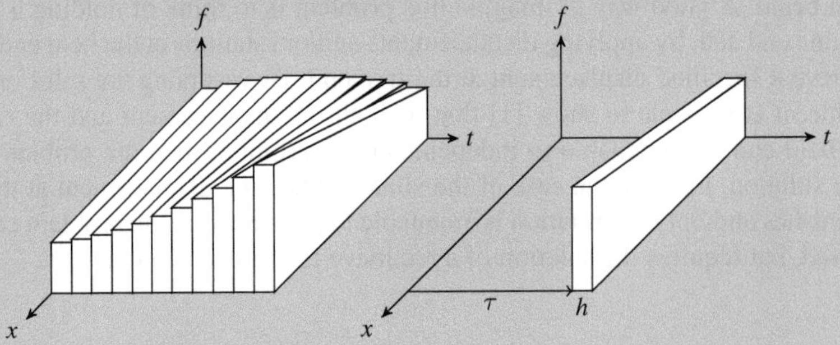

If one of these slices were acting alone starting at time τ for the duration h and then disappearing, it would leave the system, initially at rest, with a certain velocity distribution. The effect of having had the force acting for a short duration and then removed is, one claims, the same as not having had the force at all but applying instead that velocity distribution as an 'initial' condition. A glance at the PDE is convincing enough to conclude that the velocity in question is, to a first degree of approximation, $v(x, \tau + h) = f(x, \tau)h$. A more particular argument is suggested in Exercise 8.13. Since our system is linear, it abides by the principle of superposition. The combined effect of the individual slices of force is, therefore, the sum of the effects produced by the corresponding homogeneous problems with the equivalent initial conditions of velocity. In the limit, as $h \to 0$, this sum tends to the integral propounded by Duhamel's principle.

$$u(x, \tau) = 0 \qquad u_t(x, \tau) = f(x, \tau). \tag{8.67}$$

According to Duhamel's principle, the solution of the inhomogeneous equation (8.66) is given by

$$u(x, t) = \int_0^t \bar{u}(x, t; \tau)d\tau. \tag{8.68}$$

Notice the subtlety that the expression $\bar{u}(x, t; t)$ implies that t denotes the reference time and, therefore, tautologically,

$$\bar{u}(x, t; t) = 0, \tag{8.69}$$

identically for all t. Similarly,

$$\bar{u}_t(x, t; t) = f(x, t). \tag{8.70}$$

To verify that the expression (8.68) actually satisfies the PDE (8.66), we evaluate

$$u_{xx}(x, t) = \int_0^t \bar{u}_{xx}(x, t; \tau) d\tau. \tag{8.71}$$

Moreover,

$$u_t(x, t) = \bar{u}(x, t; t) + \int_0^t \bar{u}_t(x, t; \tau) d\tau = \int_0^t \bar{u}_t(x, t; \tau) d\tau, \tag{8.72}$$

where we have enforced (8.69). Finally, by virtue of (8.70),

$$u_{tt}(x, t) = \bar{u}_t(x, t; t) + \int_0^t \bar{u}_{tt}(x, t; \tau) d\tau = f(x, t) + \int_0^t \bar{u}_{tt}(x, t; \tau) d\tau. \tag{8.73}$$

The result follows directly from Eqs. (8.71) and (8.73).

Example 8.1 Solve the Cauchy problem

$$u_{tt} - c^2 u_{xx} = e^x, \qquad t \geq 0, \tag{8.74}$$

with the initial conditions

$$u(x, 0) = u_t(x, 0) = 0. \tag{8.75}$$

Solution: We need to find the solution $\bar{u}(x, t; \tau)$ of the homogeneous problem with 'initial' conditions

$$u(x, \tau) = 0 \qquad u_t(x, \tau) = e^x. \tag{8.76}$$

Using d'Alembert's solution according to Fig. 8.14, we obtain it as

$$\bar{u}(x, t; \tau) = \frac{1}{2c} \int_{x-c(t-\tau)}^{x+c(t-\tau)} e^z dz = -\frac{e^x}{c} \sinh c(t - \tau) \qquad t \geq \tau. \tag{8.77}$$

According to Duhamel's principle the answer is

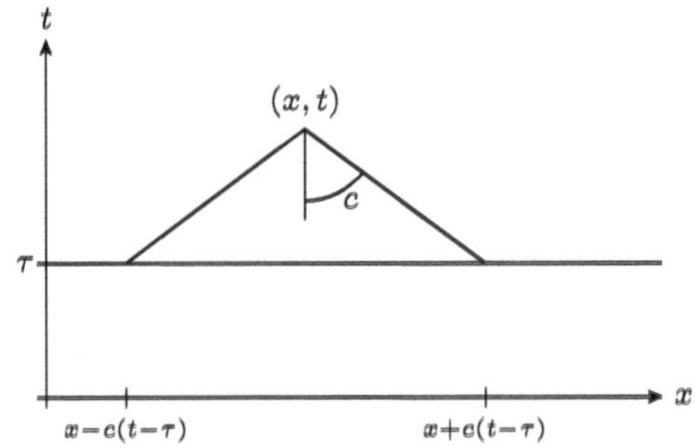

Fig. 8.14 Example of Duhamel's principle

$$u(x,t) = \int_0^t \bar{u}(x,t;\tau)d\tau = -\frac{e^x}{c}\int_0^t \sinh c(t-\tau)d\tau$$
$$= -\frac{e^x}{c^2}\left[\cosh c(t-\tau)\right]_0^t = \frac{e^x}{c^2}(\cosh ct - 1). \tag{8.78}$$

Exercises

Exercise 8.1 Make sure that you can reproduce and reason through all the steps leading to Eq. (8.13). In particular, discuss the interpretation of the two waves just introduced. Which of the two functions f, g represents the advancing wave?

Exercise 8.2 Show in detail that the solution (8.43) coincides with (8.34) in the original domain.

Exercise 8.3 (a) Use the extension technique to obtain and analyze the solution of the semi-infinite string when the left end of the string is free to move but is forced to preserve a zero slope at all times. Does the forward wave in the upper domain become the reflection of a backward wave bouncing against the support? If so, does inversion occur? (b) In terms of characteristics, place a characteristic diamond $ABCD$ with point A at the origin and point C directly above it. Show that, for the given boundary condition, $u_C = 2u_B - u_A$.

Exercise 8.4 A violin string of length L, supported at both ends, is released from rest. The initial displacement is known within a maximum point-wise error ε. Show that the same uncertainty is to be expected at any subsequent time. Hint: use the triangle inequality.

Exercise 8.5 A piano string of length L, supported at both ends, is struck at its straight configuration. The imposed initial velocity is known within a maximum point-wise error ε. Show that the point-wise uncertainty in the ensuing displacement is expected to grow linearly as time goes on. Comment on the implications of this result on any numerical method.

Exercise 8.6 Show that Eq. (8.57) implies that at a time equal to half the period, the displacements are the inverted spatial mirror image of the initial displacements, regardless of the initial velocities.

Exercise 8.7 Obtain the time periodicity of the solution strictly from a geometric analysis of the characteristics. By the same method, obtain the result of Exercise 8.6.

Exercise 8.8 Show that for the case of a free end that maintains a zero slope while the other end is fixed, the resulting motion is periodic in time. What is the period? What is the shape after half a period?

Exercise 8.9 A string of a grand piano is made of steel (density = 7800 kg/m3) and has a diameter of 0.5 mm. The string is subjected to a tension of 600 N and placed between two supports 0.3 m apart. The impact of a hammer at time $t = 0$ can be approximated by assuming a zero displacement function and a velocity given by the function

$$v_0(x) = \begin{cases} 0 & 0 \leq x < 0.18\,\mathrm{m} \\ 4\,\mathrm{m/s} & 0.18\,\mathrm{m} \leq x \leq 0.20\,\mathrm{m} \\ 0 & 0.20\,\mathrm{m} < x < 0.30\,\mathrm{m} \end{cases}$$

The origin of coordinates has been placed at one of the supports. Find the displacement of the string after 0.15 ms at the point x = 0.10 m. At the same instant, indicate those portions of the string (if any) experiencing a zero velocity. What is the period of the motion?

Exercise 8.10 Write a computer code to handle the procedure described in Sect. 8.6.1 for any given boundary displacements and initial conditions. Apply it to the case of fixed ends and verify the time periodicity of the solution.

Exercise 8.11 Write a computer code to carry out the algorithm described in Sect. 8.7. Run the program with the initial condition $u_0(x) = \sin \pi x/L$. Check the resulting shape as time goes on. Use various values of the ratio $\alpha/c < 1$.

Exercise 8.12 A Slinky toy has been idealized (not very realistically) as an elastic bar of length $L = 1$ m. This may happen when a child plastically deforms the slinky to that lamentable state so that it can no longer be enjoyed. For small additional deformations it still behaves elastically. The product EA of the modulus of elasticity times the equivalent cross-sectional area is estimated at $EA = 1N$ and the mass per unit length is $\rho A = 0.2$ kg/m. Placing the toy on a (frictionless) horizontal table to exclude gravity effects, determine the motion to be applied on one end so that the resulting displacements on the other end are given by the function $0.05(1 - \cos \omega t)H(t)$, where $H(t)$ is the Heaviside step function. Plot the solution for various values of ω in the range $1 < \omega < 15$. What happens when $\omega = 0.5\pi\sqrt{5}\,\mathrm{s}^{-1}$ or any of its odd multiples? Explain.

Exercise 8.13 (*Duhamel's principle unraveled*) Each length element of the vibrating string can ultimately be regarded as a mass attached to a spring of stiffness k somehow representing the restoring (elastic) forces of the string. Consider this mass-spring system at rest up to a time $t = \tau$ and then subjected to a force of intensity $F(\tau)$ acting for a short interval of time h and then removed. Since the spring is at rest, in the small interval of time h it will undergo a negligible displacement (of the order of h^2) while the velocity, according to Newton's second law, will undergo an increase of $F(\tau)h/m$, where m is the mass. Subject the system to an initial (at time τ) velocity $F(\tau)h/m$ and solve the homogeneous spring-mass equation $m\ddot{x} + kx = 0$ for all subsequent times. The result of this step should be

$$\bar{x}(t;\tau) = \sqrt{\frac{m}{k}}\frac{F(\tau)h}{m}\sin\left(\sqrt{\frac{k}{m}}(t-\tau)\right) \qquad t \geq \tau.$$

The superposition of all these solutions is, in the limit,

$$x(t) = \lim_{h\to 0}\sum \bar{x}(t;\tau) = \int_0^t \sqrt{\frac{m}{k}}\frac{F(\tau)}{m}\sin\left(\sqrt{\frac{k}{m}}(t-\tau)\right)d\tau.$$

Verify that this expression satisfies the differential equation $m\ddot{x} + kx = F(t)$ with zero initial conditions. Carefully distinguish between the variables t and τ and, when differentiating with respect to t, observe that it appears both in the integrand and in the upper limit of the integral. A simpler example is provided by the particular case $F(t) = \text{constant}$.

References

1. Epstein M (2013), Notes on the flexible manipulator. arXiv:1312.2912
2. Garabedian PR (1964) Partial differential equations. Wiley, New York
3. John F (1982) Partial differential equations. Springer, Berlin
4. Knobel R (2000) An introduction to the mathematical theory of waves. American Mathematical Society, Providence
5. Sneddon IN (1957) Elements of partial differential equations. McGraw-Hill, Republished by Dover, 2006
6. Sobolev SL (1989) Partial differential equations of mathematical physics. Dover, New York

Chapter 9
Standing Waves and Separation of Variables

In spite of being limited to the solution of only certain types of PDEs, the method of separation of variables provides often an avenue of approach to large classes of problems and affords important physical insights. An example of this kind is provided by the analysis of vibration problems in Engineering. The separation of variables in this case results in the resolution of a hyperbolic problem into a series of elliptic problems. The same idea will be applied in another chapter to resolve a parabolic equation in a similar manner. One of the main by-products of the method is the appearance of a usually discrete spectrum of natural properties acting as a natural signature of the system. This feature is particularly manifest in diverse applications, from musical acoustics to Quantum Mechanics.

9.1 Introduction

We have witnessed, in Chap. 8, the ability of the d'Alembert decomposition to provide solutions of the wave equation under a variety of boundary conditions. An alternative procedure to handle these and other problems can be found in the method of *separation of variables*. Among the attractive features of this method, we can mention: (i) the physical meaning that, within a given context, can usually be attributed to various elements of the procedure; (ii) the possibility to attack problems in more than one spatial dimension; and (iii) the applicability of the method beyond the realm of hyperbolic equations to linear equations of other types.[1] Nevertheless, it is important to remark that the method of separation of variables is not universally applicable. It is just one of the many techniques at our disposal to attempt to find solutions to linear

[1] Historically, in fact, the method was discovered by Jean Baptiste Joseph Fourier (1768–1830) in his celebrated book *Théorie Analytique de la Chaleur*. Thus, its very first application lies within the realm of parabolic equations.

M. Epstein, *Partial Differential Equations*, Mathematical Engineering,
DOI 10.1007/978-3-319-55212-5_9

PDEs. We will first present the method within the context of the wave equation and then, after introducing the concepts of eigenvectors and eigenvalues of differential operators, we will proceed to discuss some applications of these ideas.

9.2 A Short Review of the Discrete Case

It may be a good idea at this point to review briefly the theory of vibrations of structures with a finite number of degrees of freedom. From the mathematical point of view, as we know, these structures are governed by systems of ordinary (rather than partial) differential equations. Nevertheless, when it comes to the method of separation of variables and its consequences, there exist so many commonalities between the discrete and the continuous case that to omit their mention would constitute a callous disregard for our ability to draw intellectual parallels. Moreover, numerical methods for the solution of continuous dynamical systems usually consist of some technique of *discretization*, whereby inertia and stiffness properties are lumped in a finite number of points, thus resulting in an approximation by means of a system with a finite number of degrees of freedom.

A system of a finite number of masses interconnected by means of linear elastic springs[2] moves in space, in the absence of external forces, according to a system of linear ODEs that can be written as

$$\mathbf{K}\,\mathbf{u} + \mathbf{M}\,\ddot{\mathbf{u}} = 0. \tag{9.1}$$

In this equation $\mathbf{K}$ and $\mathbf{M}$ denote, respectively, the stiffness and mass (square) matrices of the system, while $\mathbf{u}$ is the vector of kinematic degrees of freedom (whose number n, therefore, determines the order of the matrices involved). Superimposed dots denote derivatives with respect to the time variable t. The stiffness matrix $\mathbf{K}$ is symmetric and positive semi-definite and the mass matrix $\mathbf{M}$ is symmetric and positive definite. We recall that a square matrix $\mathbf{K}$ is said to be *positive semi-definite* if for all non-vanishing vectors $\mathbf{m}$ (of the appropriate order) the following inequality is satisfied

$$\mathbf{m}^T\,\mathbf{K}\,\mathbf{m} \geq 0. \tag{9.2}$$

A square matrix is *positive definite*, on the other hand, if the strict inequality applies (that is $>$, rather than $\geq$) in Eq. (9.2). From a geometric point of view, one may say, somewhat loosely, that the vector obtained by applying the matrix to any non-zero vector never forms an obtuse angle with the original vector. In the case of positive definiteness, the angle is always acute. In the case of the stiffness and mass

[2]The linearity of the stiffness properties, which results in the linearity of the equations of motion, is either an inherent property of the system or, alternatively, the consequence of assuming that the displacements of the system are very small in some precise sense.

matrices we are discussing, these properties are consequences of reasonable physical behaviour.

A solution of the equations of motion (9.1) is a vector $\mathbf{u} = \mathbf{u}(t)$ which satisfies Eq. (9.1) identically for a given time interval. To pin down a solution, we need to specify initial conditions of position and velocity (at time $t = 0$, say). If we disregard for the moment these initial conditions, we may ask ourselves the following question: Are there any solutions that preserve their shape as time goes on? To explain this property somewhat differently, assume that as time goes on we are taking snapshots of the motion of the system. After each snapshot we record the ratios between the values of the degrees of freedom (assuming that they are not all zero at that particular time). We can call the collection of these ratios the *shape* of the present state of the system. The motion will be *shape-preserving* if its shape (namely the very collection of those ratios) is the same in each snapshot, no matter when it was taken. In other words, there is a *synchronicity* of the motion: all degrees of freedom vanish simultaneously and move, each in its own direction, in the same proportion until perhaps they all simultaneously attain an extreme value at zero speed, whereby they begin to move in the opposite direction. Now, we ask, what would the mathematical form of such a solution (if it exists) be? The answer is quite simply the following

$$\mathbf{u}(t) = \mathbf{U}\, f(t). \tag{9.3}$$

In this equation, the constant (i.e., time-independent) vector $\mathbf{U}$ represents precisely the shape of the synchronous motion, whereas the function $f(t)$ describes its evolution in time. As time goes on, the shape is preserved and the only change is in the value of the amplitude, as encapsulated in the function $f(t)$.

It is a rather remarkable fact that, under such mild assumptions, the information supplied is sufficient to calculate both the shape vector $\mathbf{U}$ and the time-evolution function $f(t)$ for all possible synchronous motions of the system. At the outset, it is important to realize that these quantities are determined only up to an arbitrary multiplicative factor, since the linearity and homogeneity of the system imply that a multiple of any solution is itself a solution.

To obtain all possible shape-preserving motions, we introduce Eq. (9.3) into (9.1) and immediately obtain

$$\mathbf{K}\,\mathbf{U}\, f + \mathbf{M}\,\mathbf{U}\, \ddot{f} = \mathbf{0}. \tag{9.4}$$

Dividing through by f, which is certainly not identically zero if we are looking for a non-trivial solution, we obtain

$$\mathbf{K}\,\mathbf{U} = -\mathbf{M}\,\mathbf{U}\, \frac{\ddot{f}}{f}. \tag{9.5}$$

By construction, the left-hand side of this equation is independent of time, and so therefore must the right-hand side be. We conclude that the ratio $\ddot{f}/f$ must be constant! To determine the possible sign of this constant, we have at our disposal the (semi-) positive definiteness of the matrices. Multiplying Eq. (9.5) to the left by the transpose of the shape vector (which certainly cannot be the zero vector, since we are only interested in non-trivial solutions), we obtain

$$\frac{\ddot{f}}{f} = -\frac{\mathbf{U}^T \mathbf{K}\, \mathbf{U}}{\mathbf{U}^T \mathbf{M}\, \mathbf{U}} \leq 0. \tag{9.6}$$

Let us investigate the case in which the equal sign holds. Notice that this is only possible because of the semi-positive definiteness of the stiffness matrix. If this matrix were positive definite, the ratio would certainly be negative. So, if the ratio is actually zero, we obtain that the second derivative of f must vanish identically, which means that the system is moving with a linearly growing amplitude. From the physical point of view, this motion corresponds to the degrees of freedom of the system as a rigid entity.[3] In other words, for a system which has been properly supported against rigid-body motions this situation cannot occur (and, in this case, the stiffness matrix is positive definite). In any case, we can write

$$\frac{\ddot{f}}{f} = -\omega^2. \tag{9.7}$$

The scalar $\omega \geq 0$ is called a *natural frequency* of the system. Can it be arbitrary? We will now discover (as, in some way, Pythagoras is said to have observed 25 centuries ago) that a linear dynamical system allows only a finite number of natural frequencies, and that these frequencies are completely determined by the stiffness and mass properties of the system. Moreover, to each of these frequencies a specific shape (or shapes) can be associated, also determined by the stiffness and inertia properties. Each of these shapes is called a *normal mode* of the system. These remarkable facts (really, things cannot get much more beautiful than this, can they?) can be generalized for continuous systems, as we shall see, thus forming the basis of musical acoustics and quantum mechanics, among other disciplines.

Introducing Eq. (9.7) into (9.5) we get

$$\left(\mathbf{K} - \omega^2\, \mathbf{M}\right)\, \mathbf{U} = \mathbf{0}. \tag{9.8}$$

This is a homogeneous system of linear algebraic equations. If the coefficient matrix $\mathbf{K} - \omega^2\, \mathbf{M}$ were non-singular, the solution of this system would be unique, thus implying that $\mathbf{U} = \mathbf{0}$, a solution which we discard (the trivial solution). Therefore, a non-trivial solution can only exist if the determinant of the coefficient matrix vanishes. We are thus led naturally to a generalized eigenvalue problem for a symmetric real matrix weighted by another matrix of the same kind. The solution of the (scalar) determinant equation

$$\det\left(\mathbf{K} - \omega^2\, \mathbf{M}\right) = 0, \tag{9.9}$$

[3] It is best here to think of the case of small rigid-body motions only.

also called the *characteristic equation*, requires the calculation of the roots of a polynomial whose degree equals the number n of degrees of freedom of the system. According to the fundamental theorem of algebra, this equation has exactly n roots $\omega_1^2, \ldots, \omega_n^2$ (some of which may be repeated). In general, some or all of these roots may be complex, but for the case of symmetric real matrices they are guaranteed to be all real. Moreover, the (semi-) positive definiteness of the matrices involved guarantees that the roots are non-negative (as we have already determined). Finally, if $\omega_i^2 \neq \omega_j^2$ and if $\mathbf{U}_i, \mathbf{U}_j$ are respectively corresponding normal modes, the following (weighted) *orthogonality condition* must be satisfied:

$$\mathbf{U}_i^T \mathbf{M}\, \mathbf{U}_j = 0. \tag{9.10}$$

In the mathematical context, each root ω_i is a (generalized) eigenvalue of the stiffness matrix weighted by the mass matrix, and the corresponding normal mode is a (generalized) eigenvector. Each of the above-mentioned properties (and some more) can be proved without too much mathematical effort.[4] To summarize the main result, we can establish that, given a linear dynamical system with n degrees of freedom, characterized by Eq. (9.1), there exists a (not necessarily unique) set of n (weighted) orthonormal eigenvectors (or normal modes of vibration) of the system. By orthonormality we mean that

$$\mathbf{U}_i^T \mathbf{M}\, \mathbf{U}_j = \delta_{ij} \qquad i, j = 1, \ldots, n, \tag{9.11}$$

where δ_{ij} is the *Kronecker symbol*, equal to 1 if $i = j$ and vanishing otherwise. Being linearly independent, these vectors can be used as a basis of the vector space $\mathbb{R}^n$.

Notice that a shape-preserving motion is necessarily of the form

$$\mathbf{u}(t) = \mathbf{U}_i\ f_i(t), \tag{9.12}$$

where the function $f_i(t)$ is a solution of Eq. (9.7) with $\omega = \omega_i$. There are clearly two possibilities. Either $\omega_i = 0$, in which case we go back to the rigid-body motion at constant speed, or $\omega_i > 0$, in which case we obtain (for some arbitrary constants A_i, B_i)

$$f_i = A_i\ \cos(\omega_i t) + B_i \sin(\omega_i t). \tag{9.13}$$

The time dependence of the (non-rigid) normal modes is, therefore, harmonic.

To complete our review of the treatment of discrete systems, we will show how, when some external forces $\mathbf{f}$ are applied in correspondence with the degrees of freedom, or when some specific initial conditions are prescribed, the solution can be represented in terms of the normal modes of the system. These concepts will reappear in a more general form in the case of continuous systems.

Let the discrete system, still in the absence of external forces, be subjected to the initial conditions

[4] An excellent reference is [2].

$$\mathbf{u}(0) = \mathbf{u}_0 \qquad \dot{\mathbf{u}}(0) = \mathbf{v}_0. \tag{9.14}$$

Since our system is linear, the principle of superposition applies, as can be easily verified. What we mean by this is that the sum of any two solutions of Eq. (9.1) is also a solution. For this reason, we attempt to represent the solution $\mathbf{u}(t)$ corresponding to the initial conditions (9.14) as a sum of independent shape-preserving motions of the form (9.12). We are setting, therefore,

$$\mathbf{u}(t) = \sum_{l=1}^{n} \mathbf{U}_l\ f_l(t) = \sum_{l=1}^{n} \mathbf{U}_l\ (A_l \cos(w_l t) + B_l \sin(\omega_l t))\,, \tag{9.15}$$

where, for definiteness, we have assumed that there are no zero eigenvalues (i.e., all rigid motions are prevented).[5]

The constants A_l, B_l will be adjusted, if possible, to satisfy the initial conditions. Multiplying Eq. (9.14) to the left by $\mathbf{U}_k^T\mathbf{M}$ and invoking the orthonormality condition (9.11), yields

$$\mathbf{U}_k^T\mathbf{M}\ \mathbf{u}(t) = A_k \cos(w_l t) + B_k \sin(\omega_l t). \tag{9.16}$$

Enforcing the initial conditions (9.14), we immediately obtain

$$A_k = \mathbf{U}_k^T\mathbf{M}\ \mathbf{u}_0, \tag{9.17}$$

$$\omega_k\ B_k = \mathbf{U}_k^T\mathbf{M}\ \mathbf{v}_0. \tag{9.18}$$

This solves the initial-value problem completely.[6]

Assume now that, in addition to the initial conditions, some external forces $\mathbf{f}$ are specified in correspondence with the degrees of freedom of the system. The equations of motion become now

$$\mathbf{K}\ \mathbf{u} + \mathbf{M}\ \ddot{\mathbf{u}} = \mathbf{f}. \tag{9.19}$$

We limit our analysis to the case in which these external forces are *harmonic*, for example of the form

$$\mathbf{f} = \mathbf{f}_0\ \sin(\omega t). \tag{9.20}$$

We look only for a *particular solution* $\mathbf{u}_p(t)$ of Eq. (9.19), since the general solution of the homogeneous equation is already available via the previous treatment. We try a solution of the form

$$\mathbf{u}_p(t) = \mathbf{U}_p\ \sin(\omega t). \tag{9.21}$$

[5]The treatment for the general case is identical, except for the fact that normal modes of the form $A_l + B_l t$ must be included.

[6]In Eq. (9.18) the summation convention does not apply.

Substitution into Eq. (9.19) yields

$$\left(\mathbf{K} - \omega^2 \, \mathbf{M}\right) \, \mathbf{U}_p = \mathbf{f}_0. \tag{9.22}$$

We conclude that, except in the case in which the frequency of the external load happens to coincide with one of the natural frequencies of the system, a particular solution of the form (9.21) is uniquely determined. The exceptional case is called *resonance* and it results in steadily increasing amplitudes of the response with the consequent disastrous effects. A nice exercise is to express the vector of the external forces in terms of the eigenvector basis and then determine the components of the particular solution in the same basis one by one. The case of a general (not necessarily harmonic) periodic force can also be handled by similar methods, but it is best treated together with the continuous case.

9.3 Shape-Preserving Motions of the Vibrating String

Although we intend to deal with more general situations, in this section we will devote our attention to the wave equation that has already occupied us in Chap. 8, namely,

$$u_{tt} = c^2 \, u_{xx}, \tag{9.23}$$

with the boundary conditions

$$u(0, t) = u(L, t) = 0 \qquad t \geq 0. \tag{9.24}$$

For now, we do not specify any initial conditions. As we know, Eqs. (9.23) and (9.24) describe the small transverse deflections of a uniform string of length L, supported at its ends, in the absence of external loading. A shape-preserving (or synchronous) motion is a solution of this equation of the form

$$u(x, t) = U(x) \; f(t). \tag{9.25}$$

In this equation, $U(x)$ represents the shape that is preserved as time goes on. A solution of this type is sometimes also called a *standing wave*. The method used to find a standing wave solution is justifiably called *separation of variables*. To see whether there are standing-wave solutions of the wave equation, we substitute the assumption (9.25) in the wave equation (9.23) and obtain

$$U(x) \; \ddot{f}(t) = c^2 \; U''(x) \; f(t). \tag{9.26}$$

We are adopting the standard notation for time and space derivatives of functions of one variable. Following the lead of the treatment of discrete systems, we now isolate (separate) all the time-dependent functions to one side of the equation and get

$$\frac{\ddot{f}(t)}{f(t)} = c^2 \frac{U''(x)}{U(x)}. \tag{9.27}$$

We conclude that each of the sides of this equation must be a constant, since a function of one variable cannot possibly be identical to a function of a different independent variable. This constant may, in principle, be of any sign. Anticipating the final result, we will presently assume that it is negative. Thus, we write

$$\frac{\ddot{f}(t)}{f(t)} = c^2 \frac{U''(x)}{U(x)} = -\omega^2. \tag{9.28}$$

We conclude, therefore, that the time-dependence of our shape-preserving candidate is harmonic, just as in the discrete case, that is,

$$f(t) = A\ \cos(\omega t) + B\ \sin(\omega t). \tag{9.29}$$

On the other hand, Eq. (9.28) also implies that (since c is a constant) the shape itself is harmonic, viz.,

$$U(x) = C\ \cos\left(\frac{\omega x}{c}\right) + D\ \sin\left(\frac{\omega x}{c}\right). \tag{9.30}$$

We still need to satisfy the boundary conditions. It follows form Eqs. (9.24) and (9.25) that, regardless of the value of ω, the boundary condition at $x = 0$ implies that

$$C = 0. \tag{9.31}$$

The boundary condition at the other end results in

$$D\ \sin\left(\frac{\omega\, L}{c}\right) = 0. \tag{9.32}$$

Since we are looking for a non-trivial solution, we must discard the possibility $D = 0$. We conclude that non-trivial solutions do exist and that they exist only for very particular values of ω, namely, those values that render the sine function zero in expression (9.32). These values (again excluding the one leading to the trivial solution) are precisely

$$\omega_k = \frac{k\ \pi\ c}{L} \qquad k = 1, 2, \ldots \tag{9.33}$$

We have thus obtained the surprising result that there exists an infinite, but discrete, *spectrum of natural frequencies* of the vibrating string corresponding to shape-preserving vibrations. The corresponding shapes, or *normal modes*, are sinusoidal functions whose half-periods are exact integer divisors of the length of the string. The fact that in this case the frequencies of the oscillations turned out to be exact multiples of each other, is the physical basis of musical aesthetics (at least until now …).

Notice that our assumption that the constant in Eq. (9.28) had to be negative is now amply justified. Had we assumed a non-negative constant, we would have been unable to satisfy both boundary conditions. In the case of the discrete system, there are no boundary conditions and the selection of natural frequencies is entirely based on the fact that the determinant (characteristic) equation, being polynomial, has a finite number of roots. In the continuous case, the selection of frequencies is mediated by the boundary conditions. By extension, the natural frequencies of the continuous case are also called *eigenvalues* of the corresponding differential operator and the normal modes of vibration are its *eigenvectors*.[7] Putting back together the spatial and temporal parts, we can express any shape-preserving solution in the form

$$u_k(x,t) = \sin\left(\frac{\omega_k x}{c}\right)(A_k\ \cos(\omega_k t) + B_k\ \sin(\omega_k t)). \tag{9.34}$$

Just as in the discrete case, the normal modes of vibration satisfy an *orthogonality* condition. Indeed, consider two different natural frequencies $\omega_i \neq \omega_j$ and the corresponding normal modes

$$U_i = \sin\left(\frac{\omega_i x}{c}\right) \qquad U_j = \sin\left(\frac{\omega_j x}{c}\right) \tag{9.35}$$

Using the trigonometric identity

$$2\ \sin\alpha\ \sin\beta = \cos(\alpha-\beta) - \cos(\alpha+\beta), \tag{9.36}$$

it is not difficult to obtain the orthogonality condition

$$\int_0^L U_i\ U_j\ dx = \int_0^L \sin\left(\frac{i\pi x}{L}\right)\ \cos\left(\frac{j\pi\ x}{L}\right)\ dx = \frac{L}{2}\ \delta_{ij}. \tag{9.37}$$

The integration of the product of two functions over the length of the domain plays, therefore, the role of a dot product in the space of functions, as we shall see later again.

[7] We content ourselves with pointing out these similarities. In fact these similarities run even deeper, particularly when we regard the underlying differential equations as linear operators on an infinite-dimensional vector space of functions, just as a matrix is a linear operator on a finite-dimensional space of vectors.

9.4 Solving Initial-Boundary Value Problems by Separation of Variables

We have found all the possible shape-preserving solutions of the one-dimensional wave equation. The method of separation of variables (which is precisely the mathematical expression of shape preservation) provided us with a harmonic time dependence, while the space dependence was settled by finding all the non-trivial solutions of the ODE

$$U''(x) = -\frac{\omega^2}{c^2}\, U(x). \tag{9.38}$$

It is interesting to compare this equation with its counterpart for the discrete case, Eq. (9.8). We start by noticing that even if the density of the string had been a function of position, the separation of variables would have resulted in a harmonic time dependence, while Eq. (9.38) would have included the variable density. We will study this case in some detail later, but we remark at this point that this variable density would have been the counterpart of the mass matrix of the discrete case.

In the discrete case, we were in possession of a *linear operator* **K** which produced vectors out of vectors (forces out of displacements, say). Now we have a linear operator on a *space of functions*. Functions defined over the same domain can be added together (point-wise) and multiplied by scalars (also point-wise). These operations endow a space of functions of a given class (say C^∞) with the structure of a vector space. The operator on the left-hand side of Eq. (9.38) consists of taking the second derivative of a function to produce another function. This operation is clearly linear (the derivative of a sum of functions is equal to the sum of the derivatives, and so on). The main difference between the discrete and the continuous case is that the space of functions under consideration is an *infinite-dimensional vector space*.

The problem posed by Eq. (9.38) can be stated as follows: Find those functions (read: 'vectors') which, when used as an input for our operator (the second derivative) result in an output proportional to the input. But this is precisely the statement of an eigenvalue problem! When we solved this problem for the discrete case, we found that there was a finite number of eigenvalues and that one can construct an orthogonal basis of eigenvectors. In the continuous case, we found that the number of eigenvalues is (countably) infinite and that the corresponding eigenvectors (which we called normal modes) are orthogonal, in the sense of Eq. (9.37).

Our claim now is that, in some precise sense, these normal modes can be understood as a basis of the infinite-dimensional space of functions under consideration. By this we mean that: (i) the normal modes are linearly independent; and (ii) every smooth enough function that vanishes at the ends of the interval under consideration can be expressed as an (infinite) linear combination of the normal modes, with coefficients to be determined uniquely for each function. In other words, we claim that 'any' function $F(x)$ defined in the interval $[0, L]$ and vanishing at its ends can be represented uniquely as a series of the form

$$F(x) = \sum_{n=1}^{\infty} D_n \, U_n(x), \tag{9.39}$$

where the numbers D_n are the *components* of the representation. This bold statement must be justified in precise mathematical terms. In this book, however, we will adopt it as an act of faith.

For the case of the wave equation that we have just considered, the normal modes are sine functions and the expansion (9.39) reduces to a particular case of the Fourier series, which we will encounter in Chap. 10. The fully-fledged Fourier series includes also terms involving the cosine function and it can be used to represent 'any' periodic function (or 'any' function defined over a finite interval that has been extended to a periodic function over the whole line). In more general cases, such as that of a non-uniform string that we will study next, the normal modes (or *eigenfunctions*) are no longer sines or cosines, but the validity of the representation (9.39) in terms of normal modes is preserved. These topics are known as Sturm–Liouville eigenvalue problems and are discussed in mathematical textbooks.[8] The more general proofs of convergence pertain to the field of functional analysis.

From the physical point of view, we may say that nature has endowed the vibrating string (and, in fact, all elastic systems) with a preferred set of shape-preserving vibrations, each of which oscillates with a definite frequency. All these *natural frequencies* constitute the *characteristic spectrum* of the system. The remarkable fact is that arbitrary periodic vibrations of the system can be expressed in terms of these natural modes of vibration. Thus, any periodic vibration of the system can be *analyzed* in terms of its *spectral components*. An intuitive grasp of these facts was perhaps somehow obtained by Pythagoras over 25 centuries ago. It is worth pointing out that the contribution of Louis de Broglie to the understanding of the quantum mechanical model of the atom runs along similar lines.

Let $F(x)$ be a function (vanishing at the ends of the interval) for which we want to find the representation (9.39). In other words, we are interested in obtaining the value of the coefficient D_k for each and every k. Multiplying both sides of (9.39) by $U_k(x)$, integrating both sides of the resulting equation over the interval $[0, L]$ and invoking the orthogonality conditions (9.37), we obtain the following surprisingly simple result

$$D_k = \frac{2}{L} \int_0^L F(x) \, U_k(x) \, dx. \tag{9.40}$$

The orthogonality conditions played a major role in the de-coupling of the formulas for the individual coefficients, just as they do in the discrete case (the components of a vector in an orthonormal basis are simply the dot products by the base vectors). There is here a subtlety, however, that we must mention. It has to do with the fact that we have assumed that the series on the right-hand side of (9.39) can be integrated term by term. Moreover, we have not specified in what precise sense the series converges to

[8] A short treatment can be found in [1], p. 291.

the given function, if it does indeed converge. These and other similar issues (which are not very difficult to understand) are outside the scope of these notes.

Let us assume that the vibrating string is subjected at time $t = 0$ to a displacement $u(x, 0) = F(x)$ and a velocity $u_t(x, 0) = G(x)$, both vanishing at the ends of the string (which remain fixed during the interval of time under consideration). These two functions can be expanded in terms of the normal modes as

$$F(x) = \sum_{n=1}^{\infty} D_n\ U_n(x), \tag{9.41}$$

$$G(x) = \sum_{n=1}^{\infty} E_n\ U_n(x). \tag{9.42}$$

The coefficients D_n, E_n can be obtained by the prescription (9.40) applied to the respective functions. We now represent the solution to the wave equation under the given initial conditions in the form

$$u(x, t) = \sum_{n=1}^{\infty} U_n(x) \left(A_n\ \cos(\omega_n t) + B_n\ \sin(\omega_n t)\right). \tag{9.43}$$

Put differently, we expand the solution in terms of the normal modes, each oscillating at its characteristic frequency. Our task is to determine all the constants A_n, B_n. At the initial time, Eq. (9.43) yields

$$u(x, 0) = \sum_{n=1}^{\infty} U_n(x)\ A_n \tag{9.44}$$

and

$$u_t(x, 0) = \sum_{n=1}^{\infty} U_n(x)\ \omega_n\ B_n. \tag{9.45}$$

In obtaining this last equation we have assumed that the series can be differentiated term by term. Comparing these results with those of Eqs. (9.42) and (9.43), respectively, and recalling that the coefficients of the expansion in terms of normal modes are unique, we conclude that

$$A_n = D_n \tag{9.46}$$

and

$$B_n = \frac{E_n}{\omega_n}. \tag{9.47}$$

This completes the solution of the problem.

The method just used is based on *normal-mode superposition*. It relies on the fact that for a homogeneous linear equation with homogeneous boundary conditions, the sum of two solutions is again a solution, and so is the product of any solution by a constant. In engineering applications (for example in structural engineering) this fact constitutes the so-called *principle of superposition*. It is not a principle of nature, but rather a property of linear operators.

If we consider a non-homogeneous situation (either because there is an external production, such as a force acting on a structure, or because the boundary conditions are not homogeneous, such as a prescribed displacement or slope at the end of a beam), it is easy to show (by direct substitution) that the difference between any two solutions of the non-homogeneous case is necessarily a solution of the homogeneous equation with homogeneous boundary conditions. It follows that the solution of a non-homogeneous problem, with prescribed *initial* conditions, can be obtained by adding any one particular solution of the non-homogeneous problem (regardless of initial conditions) to the general solution of the homogeneous problem. The adjustable constants of the homogeneous solution can then be determined so as to satisfy the initial conditions for the sum thus obtained.

To illustrate these ideas, let us consider first the case in which the string has been loaded by means of a periodic load of the form

$$q = Q(x)\ \sin(\omega t). \tag{9.48}$$

The boundary conditions are still the vanishing of the displacements at the ends of the string. Recall that the PDE governing the situation is given by Eq. (8.5), viz.,

$$u_{tt} - c^2\, u_{xx} = q, \tag{9.49}$$

which is a non-homogeneous version of the one-dimensional wave equation. We need any particular solution of this equation, so we try a solution of the form

$$u_p(x,t) = U_p(x)\ \sin(\omega t). \tag{9.50}$$

Introducing this assumption in (9.49) and using (9.48), we obtain

$$-\,\omega^2\, U_p - c^2\, U_p'' = Q. \tag{9.51}$$

This equation should be compared with its discrete counterpart, Eq. (9.22). We have obtained a non-homogeneous ODE, which can be solved by any means. From the physical point of view, however, it is interesting to find a particular solution of this ODE by making use of a normal mode superposition. We express the solution and the loading, respectively, as

$$U_p(x) = \sum_{n=1}^{\infty} D_{pn}\, U_n, \tag{9.52}$$

$$Q(x) = \sum_{n=1}^{\infty} H_n\, U_n. \tag{9.53}$$

We now assume that the second derivative of Eq. (9.52) can be carried out term by term, and use Eq. (9.38) for the normal modes to obtain

$$U_p''(x) = \sum_{n=1}^{\infty} D_{pn}\, U_n'' = -\sum_{n=1}^{\infty} \frac{\omega_n^2}{c^2}\, D_{pn}\, U_n. \tag{9.54}$$

Combining Eqs. (9.51)–(9.54), we get

$$D_{pn} = \frac{H_n}{\omega_n^2 - \omega^2}. \tag{9.55}$$

Clearly, this solution is applicable only if the frequency of the applied harmonic force does not happen to coincide with any of the natural frequencies of the system (in which case we have the phenomenon of *resonance*, with the amplitude of the response increasing steadily in time).

If the forcing function is not harmonic (or even periodic), we can still make use of the normal mode decomposition to find a particular solution of (9.49). We assume now that the coefficients of the expansions (9.52) and (9.53) are functions of time, namely,

$$u_p(x,t) = \sum_{n=1}^{\infty} \hat{D}_n(t)\, U_n(x), \tag{9.56}$$

$$q(x,t) = \sum_{n=1}^{\infty} \hat{H}_n(t)\, U_n(x). \tag{9.57}$$

Note that the coefficients in Eq. (9.57) can be calculated, instant by instant, by the formula

$$\hat{H}_k = \frac{2}{L} \int_0^L q(x,t)\, U_k(x)\, dx. \tag{9.58}$$

Similarly, the coefficients of the yet to be determined particular solution are given by

$$\hat{D}_k = \frac{2}{L} \int_0^L u_p(x,t)\, U_k(x)\, dx. \tag{9.59}$$

Let us multiply Eq. (9.49) through by $U_k(x)$ and integrate over the string to get

$$\int_0^L \left(\frac{\partial^2 u_p}{\partial t^2} - c^2\,\frac{\partial^2 u_p}{\partial x^2}\right)\, U_k(x)\, dx = \int_0^L q\, U_k(x)\, dx. \tag{9.60}$$

Integrating by parts, and taken into consideration that the boundary terms vanish, we use Eq. (9.38) to write

$$c^2 \int_0^L \frac{\partial^2 u_p}{\partial x^2}\, U_k(x)\, dx = c^2 \int_0^L u_p\, U_k''(x)\, dx = -\omega_k^2 \int_0^L u_p\, U_k\, dx. \tag{9.61}$$

Moreover, differentiating (9.56) term by term yields

$$\frac{\partial^2 u_p}{\partial t^2} = \sum_{n=1}^{\infty} \ddot{\hat{D}}_n(t)\, U_n(x). \tag{9.62}$$

Putting all these results together, Eq. (9.60) implies that

$$\ddot{\hat{D}}_k + \omega_k^2\, \hat{D}_k = \hat{H}_k. \tag{9.63}$$

This is an ODE for the determination of the time-dependent coefficients of the particular solution of the PDE. Clearly, we only need a particular solution of this ODE. Notice that in the case in which the time dependence of Q_k happens to be harmonic, we recover the solution given by (9.55). Otherwise, we can use, for example, the Duhamel integral formula, viz.,

$$\hat{D}_k(t) = \frac{1}{\omega_k} \int_0^t \hat{H}(\tau)\, \sin(\omega_k(t-\tau))\, d\tau. \tag{9.64}$$

The complete solution of the non-homogeneous problem is thus given by

$$u(x,t) = \sum_{n=1}^{\infty} U_n(x)\, \left(\hat{D}_n + A_n\, \cos(\omega_n t) + B_n\, \sin(\omega_n t)\right). \tag{9.65}$$

The constants A_n, B_n can be adjusted to fit the initial conditions.

9.5 Shape-Preserving Motions of More General Continuous Systems

In the analysis of the vibration of general linearly elastic systems the method of separation of variables leads to the separation of the (linear) problem into a trivial temporal part and a spatial part that is governed by a PDE (or system thereof) of the elliptic type. We start with the special case of a single spatial dimension.

9.5.1 String with Variable Properties

Consider the case of a string, such as that of a musical instrument, with a smoothly varying cross section with area $A = A(x)$. In accordance with Eq. (8.5), the governing equation is

$$u_{tt} - c(x)^2\, u_{xx} = q. \tag{9.66}$$

where

$$c(x) = \sqrt{\frac{T}{\rho\, A(x)}}. \tag{9.67}$$

In other words, the mass per unit length of the string is a function of position along the string. We ask the same question as in the case of the uniform string: Are there shape preserving solutions of the homogeneous equation? If so, what is the precise shape of these solutions and how does their amplitude vary with time? We proceed in exactly the same manner as before, namely, we substitute the shape-preservation assumption

$$U(x,t) = U(x)\; f(t) \tag{9.68}$$

into (the homogeneous version of) Eq. (9.66) and obtain

$$\frac{\ddot{f}(t)}{f(t)} = c(x)^2\, \frac{U''(x)}{U(x)}. \tag{9.69}$$

Just as in the case of Eq. (9.27), we reason that both sides of this equation must necessarily be constant. Assuming this constant to be negative (for the same reasons as before, which are justified a posteriori) we obtain that *the time variation of the normal modes is again necessarily harmonic*, that is,

$$f(t) = a\; \cos(\omega t) + b\; \sin(\omega t). \tag{9.70}$$

For the shape of the normal modes, however, we obtain the ODE

$$U''(x) = -\frac{\omega^2}{c(x)^2}\, U(x). \tag{9.71}$$

If we compare this equation with the discrete counterpart (9.8), we see that the mass matrix corresponds to the variable mass density per unit length. We need now to ascertain the existence of non-trivial solutions of Eq. (9.71) satisfying the given homogeneous boundary conditions. Without entering into the subtleties of the Sturm-Liouville theory, we can convince ourselves that such solutions exist by the following intuitive argument.

The left-hand side of Eq. (9.71) is (roughly) a measure of the curvature of the solution. In this sense, Eq. (9.71) tells us that the curvature of a solution (a normal mode) is proportional to the function itself, and that the constant of proportionality must be, point by point, negative. What this means is that if, starting from the left end we assume the solution to move upwards, the curvature will be negative, and thus it will bring us downwards, and vice-versa. So, let us assume that we choose some candidate value for the natural frequency. We may be lucky and hit the other end of the string. If we don't, however, we can change the value of ω gradually until we do hit the other end. That we will always be able to do so is guaranteed by the fact that the x-axis attracts the solution towards it, according to our curvature interpretation. Moreover, once we find a value of the frequency which satisfies the condition of hitting the other boundary, we can increase it gradually until we hit the far end again. Every time we do this, we add another half wave to the shape of the solution. This hit-and-miss argument can, of course, be formalized into a proof and, perhaps more importantly, into a numerical algorithm to find the normal modes.[9]

In conclusion, although for a general non-uniform string we no longer have sinusoidal normal modes, the normal modes have a wave-like appearance. The natural frequencies, moreover, will no longer be integer multiples of each other. As a result, our ear will perceive the various frequencies as dissonant with respect to each other. That is why guitar and violin strings have a constant cross section. In the case of the drum, however, because of the two-dimensionality of the membrane, the frequencies are not integer multiples of each other even in the case of constant thickness. Hence follows the typical dissonant sound of drums and cymbals. Be that as it may, we will now prove that the normal modes (eigenfunctions) satisfy a generalized (or weighted) orthogonality condition, just as in the case of the discrete system.

Let U_m, U_n be two normal modes corresponding, respectively, to two different natural frequencies ω_m, ω_n. We have, therefore,

$$U_m''(x) = -\frac{\omega_m^2}{c(x)^2}\, U_m(x), \tag{9.72}$$

[9] See Exercise 9.6.

$$U_n''(x) = -\frac{\omega_n^2}{c(x)^2}\, U_n(x). \tag{9.73}$$

Multiplying Eq. (9.72) by U_n and Eq. (9.73) by U_m, subtracting the results and integrating over the length of the string yields

$$\int_0^L \left(U_n\, U_m'' - U_m\, U_n''\right)\, dx = -(\omega_m^2 - \omega_n^2) \int_0^L \frac{U_m\, U_n}{C(x)^2}\, dx. \tag{9.74}$$

Integrating by parts the left-hand side of this equation, however, and implementing the boundary conditions, we conclude that it must vanish. Since the natural frequencies were assumed to be different, we conclude that

$$\int_0^L \frac{U_m\, U_n}{c(x)^2}\, dx = 0. \tag{9.75}$$

This is the desired generalized orthogonality condition. Since the normal modes are determined up to a multiplicative constant, we may choose to normalize them by imposing, without any loss of generality, the extra condition

$$\int_0^L \frac{U_m\, U_n}{c(x)^2}\, dx = \delta_{mn}. \tag{9.76}$$

Given a function $F(x)$ that vanishes at the ends of the string, we can express it in terms of the normal modes as

$$F(x) = \sum_{n=1}^{\infty} D_n\, U_n(x). \tag{9.77}$$

The coefficients of this expansion can be obtained by multiplying through by $U_k/c(x)^2$ and integrating term by term. Invoking our orthogonality conditions, we obtain

$$D_k = \int_0^L \frac{F(x)\, U_k(x)}{c(x)^2}\, dx. \tag{9.78}$$

From here on, the treatment of the non-uniform string is identical in all respects to that of the uniform string, provided that one takes into consideration the new normal modes and their generalized orthogonality condition.[10]

[10] In particular, the Duhamel integral will have to be expressed differently as compared to the uniform case.

9.5.2 Beam Vibrations

The equations of motion of a Bernoulli beam, which we have already encountered in Sect. 7.1.4.1, can be reduced to the single fourth-order PDE

$$(EIu_{xx})_{xx} = -q - \rho A u_{tt}, \tag{9.79}$$

in which $u = u(x, t)$ denotes the (small) transverse displacement. The free vibrations of a beam of constant properties is, therefore, governed by the homogeneous equation

$$c^4 u_{xxxx} + u_{tt} = 0, \tag{9.80}$$

with

$$c^4 = \frac{EI}{\rho A}. \tag{9.81}$$

Setting

$$u(x, t) = U(x)\ f(t), \tag{9.82}$$

yields

$$-c^4 \frac{U''''}{U} = \frac{f''}{f} = -\omega^2. \tag{9.83}$$

As before, we conclude that the time dependence of any shape-preserving motion is necessarily harmonic, with a frequency ω to be determined from the boundary conditions of the spatial problem. Some of the cases of practical interest are obtained by assuming each one of the ends $x = 0$ or $x = L$ to be *pinned* ($U = U'' = 0$), *clamped* ($U = U' = 0$) or *free* ($U'' = U''' = 0$). The general solution of the spatial equation

$$U'''' - \gamma^4 U = 0, \tag{9.84}$$

where

$$\gamma = \frac{\sqrt{\omega}}{c}, \tag{9.85}$$

can be expressed as

$$U(x) = A \sin \gamma x + B \cos \gamma x + C \sinh \gamma x + D \cosh \gamma x. \tag{9.86}$$

The four integration constants A, B, C, D are interrelated linearly by the imposition of the 4 boundary conditions corresponding to each type of support at the two ends of the beam. The resulting linear system is homogeneous and, therefore, its nontrivial solutions are found by equating the determinant of its coefficient matrix to zero. This equation delivers the natural frequency spectrum, as shown in the example of Box 9.1.

Box 9.1 The tuning fork

The tuning fork, in use since the early eighteenth century, is a U-shaped metal bar whose clearly audible fundamental (lowest) natural frequency is used as a reference to tune musical instruments. Each tine of the fork, of length L, can be considered as a Bernoulli beam with one end clamped and the other end free. Imposition of the corresponding boundary conditions results in the homogeneous system

$$\begin{bmatrix} 0 & 1 & 0 & 1 \\ \gamma & 0 & \gamma & 0 \\ -\gamma^2 \sin \gamma L & -\gamma^2 \cos \gamma L & \gamma^2 \sinh \gamma L & \gamma^2 \cosh \gamma L \\ -\gamma^3 \cos \gamma L & \gamma^3 \sin \gamma L & \gamma^3 \cosh \gamma L & \gamma^3 \sinh \gamma L \end{bmatrix} \begin{Bmatrix} A \\ B \\ C \\ D \end{Bmatrix} = \begin{Bmatrix} 0 \\ 0 \\ 0 \\ 0 \end{Bmatrix}.$$

The determinant of the coefficient matrix is

$$\Delta = 2\gamma^6 (1 + \cos \gamma L \ \cosh \gamma L).$$

We conclude that the natural frequencies are obtained from the roots of the transcendental equation

$$\cos \gamma L = -\frac{1}{\cosh \gamma L}.$$

Written in this way, since the right-hand side decays extremely fast, the natural frequencies beyond the first and the second are obtained with very small error from $\cos \gamma L = 0$. The first and second roots are, respectively, $\gamma_1 L = 0.597\pi$ and $\gamma_2 L = 1.494\pi$. The lowest natural frequency is, therefore,

$$\omega_1 = \gamma_1^2 c^2 = \left(\frac{0.597\pi}{L}\right)^2 \sqrt{\frac{EI}{\rho A}}.$$

The dimensions can be easily calibrated to produce the usual orchestra pitch A440.

9.5.2.1 Orthogonality of the Normal Modes

Let $U_i(x)$ and $U_j(x)$ be normal modes corresponding to two distinct eigenvalues γ_i^4 and γ_j^4, respectively for some specific support conditions. We recall that these eigenfunctions are obtained, up to a multiplicative constant, from nontrivial solutions of the homogeneous system of equations for the constants of integration. Exploiting Eq. (9.84) and integrating by parts, we obtain

$$\begin{aligned} (\gamma_j^4 - \gamma_i^4) \int_0^L U_i U_j dx &= \int_0^L (U_i U_j'''' - U_j U_i'''') dx \\ &= \int_0^L (-U_i' U_j''' + U_j' U_i''') dx + \left[U_i U_j''' - U_j U_i''' \right]_0^L \\ &= \left[U_i U_j''' - U_j U_i''' - U_i' U_j'' + U_j' U_i'' \right]_0^L . \end{aligned} \tag{9.87}$$

The last expression vanishes by virtue of the boundary conditions, whence the desired orthogonality.

An important feature of eigenfunctions beyond their orthogonality is their *completeness* in the sense that every smooth enough function can be expressed as an infinite linear combination of these basic functions.[11] In particular, this feature is helpful in solving non-homogeneous problems by expressing the forcing load in terms of the eigenfunctions, as was done for the vibrating string.

9.5.3 The Vibrating Membrane

As a two-dimensional version of the vibrating string, the vibrating membrane is a thin elastic sheet, with negligible bending stiffness, stretched uniformly in a plane by means of a tension T, measured in units of force per unit length, and then placed within a rigid frame, like the skin of a drum or a trampoline. We assume that the displacements are only transversal and, whether due to the applied load or the vibrating process or possibly small deformations of the frame, they are very small when compared with the overall dimensions in the plane. We already introduced the governing equation (2.43) as

$$u_{xx} + u_{yy} = -\frac{q}{T} + \frac{\rho h}{T}\, u_{tt}, \tag{9.88}$$

where h is the thickness of the membrane and ρ the density of the material. In the absence of the external loading term q, we obtain the *two-dimensional wave equation*, which we already encountered in Sect. 7.2.4. The elimination of the time dependence leads to the elliptic equilibrium equation

$$u_{xx} + u_{yy} = -\frac{q}{T}. \tag{9.89}$$

This is the *Poisson equation*. If the right-hand side vanishes we obtain the *Laplace equation*. These equations appear in many engineering applications, including fluid mechanics, acoustics, electrostatics and gravitation.

The free vibrations of the membrane are described by the hyperbolic PDE

$$u_{xx} + u_{yy} = \frac{1}{c^2} u_{tt}, \tag{9.90}$$

with $c^2 = T/\rho h$ representing the speed of propagation of signals. In terms of the Laplacian operator introduced in Eq. (1.21), we can rewrite (9.90) more compactly as

$$\nabla^2 u = \frac{1}{c^2} u_{tt}. \tag{9.91}$$

[11] For a proof, see [1], p. 359.

A shape-preserving motion is of the form

$$u(x, y, t) = U(x, y)\ f(t), \tag{9.92}$$

which leads to the relation

$$c^2\ \frac{\nabla^2 U}{U} = \frac{f''}{f} = -\omega^2. \tag{9.93}$$

As before, we obtain for the time dependence the familiar

$$f(t) = A \sin \omega t + B \cos \omega t. \tag{9.94}$$

As far as the shape itself is concerned, we are left with the elliptic PDE

$$\nabla^2 U + \frac{\omega^2}{c^2} U = 0. \tag{9.95}$$

In other words, the natural frequencies ω, as expected, will be determined by solving the eigenvalue problem for the linear differential operator ∇^2. The selection of the frequency spectrum will depend on the shape and size of the membrane domain $\mathcal{A}$ and on the type of boundary conditions imposed, just as in the previous problems. Since the Laplacian operator is elliptic, we have yet to consider this kind of problems. We can, however, advance a few useful considerations. We start by observing that the divergence theorem (1.18) can be restated, with the appropriate interpretation, in a two-dimensional (rather than three-dimensional) Cartesian domain $\mathcal{A}$, in which case the flux is evaluated over the boundary curve $\partial\mathcal{A}$. Applying the theorem to a vector field given by the product of a scalar field ϕ time the gradient of another scalar field ψ, we obtain

$$\int_{\mathcal{A}} \nabla \cdot (\phi \nabla \psi)\ d\mathcal{A} = \int_{\partial\mathcal{A}} \phi \frac{d\psi}{dn}\ ds, \tag{9.96}$$

where n denotes the normal to the boundary curve. On the other hand, it is easy to check that

$$\nabla \cdot (\phi \nabla \psi) = \nabla \phi \cdot \nabla \psi + \phi \nabla^2 \psi. \tag{9.97}$$

Combining these results, we obtain the identity

$$\int_{\mathcal{A}} (\phi \nabla^2 \psi - \psi \nabla^2 \phi)\ d\mathcal{A} = \int_{\partial\mathcal{A}} \left(\phi \frac{d\psi}{dn} - \psi \frac{d\phi}{dn} \right) ds. \tag{9.98}$$

This elegant result has important consequences. Let $U_i(x, y)$ and $U_j(x, y)$ be eigenfunctions (that is, natural modes of vibration) corresponding to two distinct natural frequencies ω_i, ω_j, respectively. Then, emulating Eq. (9.74) or (9.87), we can write

$$\frac{1}{c^2}(\omega_j^2 - \omega_i^2)\int\limits_{\mathcal{A}} U_i U_j d\mathcal{A} = -\int\limits_{\mathcal{A}} (U_i \nabla^2 U_j - U_j \nabla^2 U_i) d\mathcal{A}$$
$$= -\int\limits_{\partial\mathcal{A}} \left(U_i \frac{dU_j}{dn} - U_j \frac{dU_i}{dn} \right) ds. \qquad (9.99)$$

What this result entails is that the orthogonality condition between eigenfunctions will be satisfied for at least two types of boundary conditions. The first type corresponds to a simple support, namely, to the vanishing of the transverse displacement. This kind of boundary condition ($U = 0$) for an elliptic operator is known as the *Dirichlet type*. The other kind of boundary condition ($dU/dn = 0$), known as the *Neumann type*, corresponds to the vanishing of the slope of the membrane.[12] In more general problems of Elasticity, the Neumann boundary condition corresponds to the specification of a surface traction. At any rate, we conclude that the eigenfunctions corresponding to different eigenvalues are orthogonal. It can also be shown that they form a complete set and that any sufficiently smooth function defined over the domain $\mathcal{A}$ can be expanded in terms of the set formed by all the eigenfunctions.

Although we will not pursue at this stage the actual solution of the elliptic equation in the general case, it turns out that a further application of the method of separation of variables will allow us to solve the problem for the case of a rectangular membrane. Indeed, let the domain $\mathcal{A}$ be the Cartesian product $[0, a] \times [0, b]$, where a and b are the lengths of the sides. Consider the case of a simply supported membrane along the whole perimeter. We try a variable separated solution of Eq. (9.95) in the form

$$U(x, y) = X(x)Y(y), \qquad (9.100)$$

which yields

$$\frac{d^2X}{dx^2} + \frac{d^2Y}{dy^2} = -\frac{\omega^2}{c^2}. \qquad (9.101)$$

This identity is only possible if each of the summands is constant, namely, after solving and imposing the boundary conditions at $x = 0$ and $y = 0$, if

$$X(x) = A\sin\lambda x \qquad Y(y) = B\sin\mu y \qquad \text{with } \lambda^2 + \mu^2 = \frac{\omega^2}{c^2}. \qquad (9.102)$$

Imposing the boundary conditions at $x = a$ and $y = b$ and discarding the trivial solution, we obtain the condition

$$\pi^2\left(\frac{m^2}{a^2} + \frac{n^2}{b^2}\right) = \frac{\omega^2}{c^2} \qquad m, n = 1, 2, 3, \ldots \qquad (9.103)$$

---

[12] As a matter of historical interest, the Neumann boundary condition is named after the German mathematician Carl Neumann (1832–1925), not to be confused with the Hungarian-American mathematician John von Neumann (1903–1957).

For any given pair m, n, the corresponding shape-preserving vibration $u_{mn} = u_{mn}(x, y, t)$ is

$$u_{mn} = \sin\frac{m\pi x}{a}\sin\frac{n\pi y}{b}\left(A\sin\pi c\sqrt{\frac{m^2}{a^2}+\frac{n^2}{b^2}}t + B\cos\pi c\sqrt{\frac{m^2}{a^2}+\frac{n^2}{b^2}}t\right). \tag{9.104}$$

From the musical point of view, we can see that the successive natural frequencies do not appear in ratios of small integers, which accounts for the blunt sound of drums in an orchestra. Moreover, for a ratio b/a equal to a rational number, we have multiple eigenvalues. For instance, in a square membrane, we have $u_{mn} = u_{nm}$ for any pair with $m \neq n$. In that case, any linear combination of u_{mn} and u_{nm} is again an eigenfunction corresponding to the same eigenvalue. The idea of an ocular tonometer based on a careful measurement of the normal modes of the eye and their sensitivity to small variations of the intra-ocular pressure is adversely affected by these considerations, since it is difficult to ascertain which normal mode of vibration has been excited by an external source.

Exercises

Exercise 9.1 Write the equations of motion of a system with 2 or 3 masses moving in the plane (or in the line) and interconnected by springs. Are there any non-zero vectors (that is, displacements) for which the equal sign in Eq. (9.2) applies? Why, or why not? What would the violation of the inequality mean for the stiffness and/or for the mass matrix in physical terms?

Exercise 9.2 For a given vector $\mathbf{f} = \mathbf{f}_0\ \sin(\omega t)$ in Eq. (9.19), express the vector $\mathbf{f}_0$ in terms of the eigenvector basis and then determine the components of the particular solution $\mathbf{U}_p$ in the same basis. What happens in this method when ω happens to coincide with one of the natural frequencies?

Exercise 9.3 Compare the solution given by Eqs. (9.43)–(9.47) with that given (for the same problem, with a slightly different notation) by the extension technique in Chap. 8. If you prefer, consider just the case in which the initial velocity is identically zero. Are these solutions the same, as they should?

Exercise 9.4 Verify, by direct substitution, that the integral (9.64) satisfies the ODE (9.63).

Exercise 9.5 A guitarist plucks a string of length L at exactly its midpoint. Let W denote the magnitude of the imposed deflection under the finger just before it is released from rest. Assuming the two halves of the string to be straight at that instant, determine the coefficients of its expression in terms of the eigenfunctions (i.e., perform a Fourier expansion). Plot the approximate shape obtained by using just a few terms of the expansion. Using the method of separation of variables, find the solution for the motion of the string. Plot it for various times within a period, using just a few terms of the approximation. Solve the same problem by the method of characteristics and compare the results for the same times. Comment on the comparison. Is any of the two solutions exact? If so, which one?

Exercise 9.6 Consider a vibrating string of constant density but with a linearly varying cross section

$$A(x) = A_0 \left(1 + 4\frac{x}{L}\right),$$

where L is the string length and A_0 is the area at the left end. Implement numerically the shooting method described in Sect. 9.5.1 to find the first few natural frequencies and the corresponding normal modes. You may use specific numerical values or introduce a non-dimensional coordinate $\xi = x/L$ and calculate the eigenvalues relative to those of a string with constant cross section A_0. The shooting routine can easily be handled with Mathematica by solving the ODE with zero displacement and, say, unit slope at the left end. Varying the coefficient containing the eigenvalue until the other end is hit and counting the number of oscillations, the solution is obtained in just a few runs.

Exercise 9.7 Show that for a simply supported (i.e., pinned-pinned) Bernoulli beam, the natural modes of vibration are the same as for the vibrating string, while the successive natural frequencies are not in the same relation.

Exercise 9.8 A uniform Bernoulli beam of length L is supported on a continuous elastic foundation, just like a beam lying on the ground. The differential equation for the static bending deflection $u = u(x)$ of this beam is $EIu'''' + ku = -q$. In this equation, k represents the stiffness of the foundation expressed in units of force per unit length of the beam. The ends of the beam are supported by means of immovable pins. Solve for the deflection of this beam under a constant distributed load q by expanding in terms of a Fourier sine series. If the dynamic (inertia) contribution were to be taken into account, what would the solution be?

Exercise 9.9 For the vibrating membrane, show that the orthogonality between normal modes corresponding to different natural frequencies is also verified by a third kind of boundary condition known as the *Robin type*. It corresponds physically to an elastic support that keeps the proportionality between the displacement and the normal slope of the membrane, namely, $U = k dU/dn$, where k is an elastic constant.

Exercise 9.10 A rectangular membrane with $a/b = 1.5$, simply supported around its perimeter in the x, y plane, is subjected to a uniformly distributed normal load p (per unit area of the membrane) and to a uniform tension T (load per unit length) in all directions. Find an approximate solution of this static problem by means of the technique of separation of variables. Give numerical values to the various constants and estimate the maximum deflection of the membrane to 3 significant digits.

References

1. Courant R, Hilbert D (1962) Methods of mathematical physics, vol I. Interscience, Wiley, New York
2. Hildebrand FB (1965) Methods of applied mathematics. Prentice Hall, Englewood Cliffs (Reprinted by Dover (1992))

Chapter 10
The Diffusion Equation

The archetypal parabolic equation is the diffusion equation, or heat equation, in one spatial dimension. Because it involves a time derivative of odd order, it is essentially irreversible in time, in sharp distinction with the wave equation. In physical terms one may say that the diffusion equation entails an arrow of time, a concept related to the Second Law of Thermodynamics. On the other hand, many of the solution techniques already developed for hyperbolic equations are also applicable for the parabolic case, and vice-versa, as will become clear in this chapter.

10.1 Physical Considerations

Many phenomena of everyday occurrence are by nature diffusive.[1] They arise, for example, as the result of sneezing, pouring milk into a cup of coffee, intravenous injection and industrial pollution. These phenomena, consisting of the spread of one substance within another, are characterized by *thermodynamic irreversibility* as the system tends to equilibrium by trying to render the concentration of the invading substance as uniform as possible. The flow of heat is also a diffusive process. A more graphical way to describe this irreversibility is to say that diffusive phenomena are characterized by an *arrow of time*. Thus, the drop of milk poured into the coffee will never collect itself again into a drop.

10.1.1 Diffusion of a Pollutant

Consider a tube of constant cross section filled with a liquid at rest (the *substrate*), in which another substance (the *pollutant*) is present with a variable concentration $g = g(x, t)$, measured in terms of mass of pollutant per unit length of tube. If this

[1] This section is largely a more detailed repetition of Sect. 2.4.2.

M. Epstein, *Partial Differential Equations*, Mathematical Engineering,
DOI 10.1007/978-3-319-55212-5_10

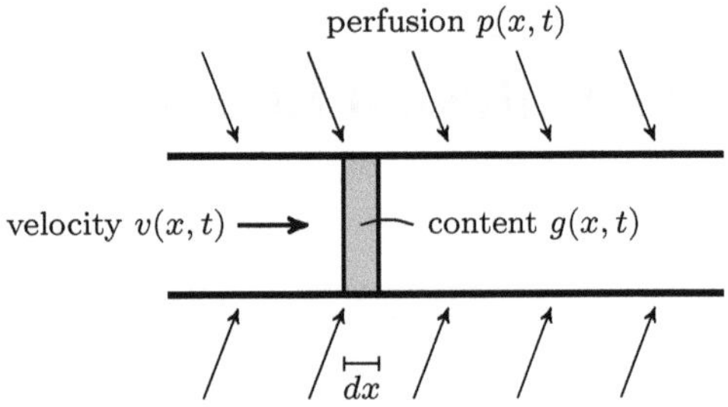

Fig. 10.1 Diffusion of a pollutant in a buried pipe

tube is embedded in a hostile environment and, if the tube wall permits it, a certain amount of pollutant, $p = p(x, t)$, may perfuse through the lateral wall per unit length and per unit time. The quantity p is usually called the *production*. We want to account for the variation in time of the amount of pollutant contained in an infinitesimal slice of width dx, as shown in Fig. 10.1.

If the pollutant were to remain at rest, this accounting would be trivial, as it would state that the change in pollutant content, namely $\frac{\partial g(x,t)}{\partial t} dx$, is entirely due to the perfusion through the lateral wall, that is, $p(x, t)\ dx$. In reality, however, the pollutant tends to move with a velocity $v(x, t)$ in the direction x of the tube axis. This motion, which is the essence of the diffusive phenomenon, results in an inflow through the left face of the slice given by $g(x, t)\ v(x, t)$, measured in mass of pollutant per unit time. Analogously, the right face of the slice, located at the spatial position $x + dx$, will witness an outflow of pollutant in the amount $g(x + dx, t)\ v(x + dx, t)$. The net contribution due to flow through the faces is, therefore, given by $g(x, t)\ v(x, t) - g(x + dx, t)\ v(x + dx, t) = -\frac{\partial (gv)}{\partial x}\ dx + O(dx^2)$, where we have assumed the quantities involved to be differentiable. Adding up the various contributions, we obtain in the limit as $dx \to 0$ the balance equation

$$\frac{\partial g}{\partial t} + \frac{\partial (gv)}{\partial x} = p. \tag{10.1}$$

To complete the physical description of the diffusion phenomenon, we need to supply a *constitutive equation* that relates the two dependent field variables v and g. In the case of diffusion of a pollutant (or, in general, a substance in small concentrations within another), it is possible to formulate a sensible, experimentally based, constitutive law directly in terms of the pollutant concentration. The most commonly used model, called *Fick's law*, states that

$$gv = -D \text{ grad } g, \tag{10.2}$$

where grad denotes the spatial gradient and the positive constant D is a property that depends on the substances involved. The minus sign in Eq. (10.2) agrees with the fact that the pollutant tends to flow in the direction of smaller concentrations.

Combining the last two equations, we obtain the second-order linear PDE

$$\frac{\partial g}{\partial t} - D\frac{\partial^2 g}{\partial x^2} = p. \tag{10.3}$$

In the absence of production we obtain the homogeneous equation

$$\frac{\partial g}{\partial t} - D\frac{\partial^2 g}{\partial x^2} = 0, \tag{10.4}$$

known as the *diffusion equation* and also as the *heat equation*. A clever statistical motivation for the diffusion equation is presented in Box 10.1.

Box 10.1 A discrete diffusion model
A plausible a priori justification of Fick's law can be made by resorting to a Statistical Mechanics argument. This kind of argument can be very fruitful in many applications by providing a heuristic link between various levels of analysis. Following a line of thought that can be regarded as a greatly simplified version of Einstein's celebrated 1905 explanation of Brownian motion, we postulate a discrete model of space and time, namely, we assume that all events take place at specific isolated sites and instants. The sites are assumed to be equally spaced along the real line according to the formula:

$$x_i = i\,h \qquad i = \ldots -3, -2, -1, 0, 1, 2, 3, \ldots,$$

where $\Delta x = h$ is the distance between neighbouring sites. Similarly, the chosen instants of time are spaced at regular intervals according to the formula

$$t_j = j\,k \qquad j = 0, 1, 2, 3, \ldots,$$

with $\Delta t = k$ being the time interval between consecutive events. We assume, moreover, that at time $t = 0$ each site is occupied by a number N_i^0 of particles and we want to establish how this discrete mechanical system will evolve thereafter, namely, we want to predict the number N_i^j of particles at the site x_i at time t_j.

In the intended physical picture we imagine that there is an underlying ground substance (coffee, say) in equilibrium and that the particles of interest (a drop of milk, say) are being constantly bombarded by collisions with the molecules making up the ground substance. Because these collisions are random, we assume that each particle has an equal probability $\beta = 0.5$ to move one space either to the right or to the left in the time interval $\Delta t = k$. Under this basic assumption, the rule of evolution of this system[a] is given by

$$N_i^{j+1} = 0.5N_{i-1}^j + 0.5N_{i+1}^j.$$

The link between the discrete model and the diffusion equation is obtained by formulating the latter as a finite-difference approximation on the assumed space-time grid. Setting $g(x_i, t_j) = N_i^j/h$ and using standard approximation formulae for first and second derivatives, we obtain

$$\frac{N_i^{j+1} - N_i^j}{k} \approx D\,\frac{N_{i-1}^j - 2N_i^j + N_{i+1}^j}{h^2}.$$

Setting $D = h^2/2k$, we recover the discrete version. The diffusion coefficient D is thus seen to be related directly to the average particle distance and the mean time between collisions.

[a] A discrete system governed by an evolution rule that determines the next state on the basis of the present state only is called a *cellular automaton*.

10.1.2 Conduction of Heat

The *First Law of Thermodynamics* asserts that for each substance there exists a function of state, called the *internal energy*, whose rate of change is balanced by the power of the external forces (or mechanical power) acting on the system plus the heating input (or thermal power), namely,

$$\frac{d\ \text{internal energy}}{dt} = \text{mechanical power} + \text{thermal power}. \tag{10.5}$$

If we consider a fixed non-deforming substrate, such as a metal wire, the mechanical power vanishes and the free energy is a function of the temperature alone. In close analogy with the diffusion case shown in Fig. 10.1, the thermal power going into a slice of width dx consists of two parts: (i) A power supply $p = p(x, t)$, measured in terms of energy per unit length and per unit time. This power is the result of sources of heat distributed throughout the length of the wire (or its lateral surface). (ii) A heat flux $q = q(x, t)$ in the direction of the axial coordinate x. This flux, due to the ability of the material to conduct heat, is measured in terms of energy per unit time.[2] Denoting by $g = g(x, t)$ the temperature field and by $u = u(x, t)$ the internal energy

[2] In the more general three-dimensional context, the production term p is measured per unit volume (rather than length) and the flux term q is measured per unit area. Since the cross section has been assumed to be constant, we did not bother to effect the formal passage to one dimension.

content per unit length, the statement of the balance of energy is expressed as

$$\frac{\partial u(x,t)}{\partial t} = p(x,t) - \frac{\partial q(x,t)}{\partial x}. \tag{10.6}$$

For many materials, a good empirical constitutive law for increments Δu in internal energy due to corresponding increments Δg in temperature is given by the linear relation

$$\Delta u = c \, \Delta g, \tag{10.7}$$

where c is a constant known as the *specific heat (capacity)*. The internal energy depends in general also on the deformation, which in our case has been ignored since the material was assumed to be rigid.

As far as the heat flux is concerned, *Fourier's Law of heat conduction* is an empirical relation valid for most materials within limited temperature ranges. It establishes that the heat flux is proportional to the gradient of the temperature. In our notation, this law is expressed as

$$q = -k \, \frac{\partial g}{\partial x}, \tag{10.8}$$

where k is the *thermal conductivity* of the material, a positive constant. The minus sign expresses the fact that heat flows spontaneously form higher to lower temperatures.

Introducing the constitutive equations (10.7) and (10.8) into the energy balance equation (10.5), we obtain

$$c\frac{\partial g}{\partial t} - k\frac{\partial^2 g}{\partial x^2} = p. \tag{10.9}$$

This equation is identical in form[3] to Eq. (10.3) governing the diffusion of one substance into another, as we studied in Sect. 10.1.1. For this reason, this equation is known both as the (non-homogeneous) *diffusion equation* and as the *heat equation*. The adjective non-homogeneous refers here to the fact that there are body sources. Thus, the equation would be called homogeneous if the right-hand side were zero. On the other hand, the material itself may have properties, such as the specific heat or the thermal conductivity, varying from point to point, in which case it is the body (rather than the equation) which would be called inhomogeneous. In deriving Eq. (10.9), in fact, it was assumed that the coefficient of thermal conductivity k was constant throughout the domain of interest. If, instead, k and/or c are functions of position (that is, if the material is inhomogeneous) Eq. (10.9) should be replaced by

$$c(x) \, \frac{\partial g}{\partial t} - \frac{\partial}{\partial x} \left(k(x) \frac{\partial g}{\partial x} \right) = p. \tag{10.10}$$

[3] With $D = k/c$.

10.2 General Remarks on the Diffusion Equation

The diffusion equation, as we already know, is of the *parabolic type*. At each point, therefore, it has a single characteristic direction, whose slope is given by

$$\frac{dt}{dx} = 0. \tag{10.11}$$

There are some similarities between the heat equation and the wave equation (and between hyperbolic and parabolic equations in general), but there are also many differences, both in interpretation and in the nature of their solutions, which we would like to point out. An important difference from the mathematical and physical points of view is that, unlike the wave equation, the heat equation is not invariant with respect to time reversal. In other words, if we were to make the change of independent variables

$$\begin{cases} \hat{x} = x \\ \hat{t} = -t \end{cases} \tag{10.12}$$

we would not obtain the same equation (as would be the case with the wave equation). From the physical point of view, this is a manifestation of the fact that the diffusion equation describes thermodynamically *irreversible* processes. If we were to run a video of a wave phenomenon backwards, we would not be able to tell whether or not we are witnessing a real phenomenon. But if we were to run backwards a movie of a diffusive phenomenon, we would immediately be able to tell that something "unphysical" is taking place: the milk already dissolved in coffee spontaneously becomes a small drop, a bar in thermal equilibrium spontaneously gets cooler at one end and warmer at the other, and so on.

A feature that the diffusion equation shares with the wave equation is that in both cases the initial value problem makes physical and mathematical sense. Thus, from the state of the system at one particular instant of time, the differential equation allows us to predict the evolution of the system for future times. But here, again, there is a significant difference when we realize that, according to Eq. (10.11), the initial curve ($t = 0$, say) is a characteristic of the PDE. What this means is that we will not be able to consider the specification of initial data along a characteristic as the exceptional case, but rather as the rule. Moreover, if we recall that characteristics are lines along which weak singularities propagate, we find that according to the heat equation these disturbances (if they could exist at all) must propagate at an infinite speed! In fact, it can be shown that, in the interior of its domain of existence, any solution of the heat equation must be of class C^∞. In other words, contrary to the wave equation, any singularity in the data at the boundary of the domain is immediately smeared out, as befits a diffusive process.

When discussing the different types of second-order equations, we remarked that if the initial data (the values of the function and of its first partial derivatives) are specified on a characteristic line they will in general contravene the PDE. In the case of hyperbolic equations, upon reconciling the initial data with the PDE we end up losing uniqueness of the solution. In order to restore it, one has to deal with the so-called *characteristic initial value problem*, whereby data have to be specified on two intersecting characteristics. In the case of the diffusion equation, on the other hand, it is clear that to reconcile the initial data on a characteristic line (t = constant) we simply have to refrain from specifying the value of the time derivative $g_t = \frac{\partial g}{\partial t}$. Indeed, by specifying the value of the function itself, and assuming that it is twice differentiable, we can obtain its second spatial derivative, and the PDE automatically delivers the value of the time derivative. We note, moreover, that the values of all subsequent partial derivatives of all orders become thus available. This means that, contrary to the case of the wave equation, the reconciliation of the initial data with the PDE does not lead to a lack of uniqueness.

10.3 Separating Variables

Without assuming any initial and/or boundary conditions, we try to see whether the method of separation of variables can give us some indication of possible solutions of the homogeneous diffusion equation. We set

$$g(x,t) = G(x)\ f(t). \tag{10.13}$$

Upon substitution in (10.4), we obtain the relation

$$\frac{\dot{f}}{f} = D\,\frac{G''}{G} = -\lambda^2, \tag{10.14}$$

with an obvious notation. Without any loss of generality, we may henceforth assume that $D = 1$, since this can always be achieved by a suitable re-scaling of the spatial variable. The choice of negative sign in the constant rightmost side of Eq. (10.14) is dictated by the reasoning that follows. Integrating first the time-dependent part, we obtain

$$f(t) = C\ e^{-\lambda^2 t}. \tag{10.15}$$

Since, as already pointed out, the diffusion equation implies an arrow of time, it is clear that, had we chosen a positive value in the right-hand side, the solution would have rapidly diverged with (increasing) time. The spatial part leads to the solution

$$G(x) = A\ \cos(\lambda x) + B\ \sin(\lambda x). \tag{10.16}$$

Absorbing the undetermined constant C within A and B, we obtain a solution of the heat equation for $\lambda \neq 0$, namely,

$$g(x,t) = (A\ \cos(\lambda x) + B\ \sin(\lambda x))\ e^{-\lambda^2 t}. \tag{10.17}$$

Note that the special choice $\lambda = 0$ yields the solution $g(x,t) = A + Bx$. This time-independent solution of the diffusion equation corresponds to a case of *steady state* (or equilibrium).

Since we are dealing with a linear homogeneous equation, any linear combination of solutions is a solution. We may, for example, choose for each value of λ some prescriptions $A = A(\lambda)$ and $B = B(\lambda)$, and form an integral (which is, after all, a limit of sums) such as

$$g(x,t) = \int_{-\infty}^{\infty} (A\,\cos(\lambda x) + B\,\sin(\lambda x))\; e^{-\lambda^2 t}\, d\lambda. \tag{10.18}$$

Provided the integral converges, this expression is a new solution of the diffusion equation. We will later exploit this fact to show how to construct a solution in this way by judiciously choosing the *spectral coefficients* $A(\lambda)$ and $B(\lambda)$ so as to match any given initial and boundary conditions.

10.4 The Maximum–Minimum Theorem and Its Consequences

Before presenting the solutions of particular initial-boundary value problems, it is instructive to dwell on some of the properties of these solutions. Are they unique? Can they sustain discontinuities? Do they depend continuously on the initial data? A particularly illuminating method to answer these questions in the case of parabolic equations is based on the maximum–minimum theorem, whose statement can and will be later interpreted in appealingly physical terms.[4]

Theorem 10.1 (Maximum–minimum theorem) *Let $\mathcal{D}$ denote a closed rectangle with sides parallel to the coordinate axes. Without loss of generality (since the heat equation is invariant under translations in both directions), we may assume that this rectangle is $[0, L] \times [0, T]$, as shown in the Fig. 10.2. Any (continuous)[5] solution of the diffusion equation (10.4) in $\mathcal{D}$ attains its maximum and minimum values either on the base ($t = 0$) of the rectangle or on one of the vertical sides ($x = 0, L$).*

Proof Let M and m denote, respectively, the maximum values[6] of g in $\mathcal{D}$ and in the union of the base and the vertical sides of $\mathcal{D}$ (that is, in $\partial\mathcal{D}$ minus the open top of the rectangle, as indicated with a thick line in Fig. 10.2). Assume the statement of

[4]This section follows [5].

[5]As already pointed out, any solution is already smooth in the interior of the domain. Continuity refers, therefore, to the data specified on (part of) the boundary.

[6]Recall that a continuous function defined over a compact (closed and bounded) subset of $\mathbb{R}^n$ attains its maximum and minimum values at one or more points of its domain.

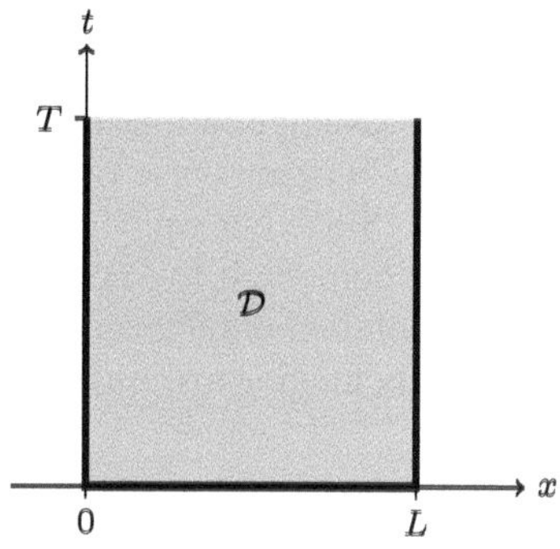

Fig. 10.2 The maximum–minimum theorem

the theorem not to be true. There exists, therefore, a point (x_0, t_0) with $0 < x_0 < L$ and $0 < t_0 \leq T$ at which g attains the value $M > m$. We construct the augmented function

$$h(x, t) = g(x, t) + \frac{M - m}{4L^2}(x - x_0)^2. \tag{10.19}$$

The reason for this construction will become apparent soon. We only remark now that the value of this function is at each point of $\mathcal{D}$ greater than or equal to the value of the solution at that point. In particular, the restriction of this function to the union of the vertical sides and the base of $\mathcal{D}$ satisfies the inequality

$$h(x, t) \leq m + \frac{M - m}{4} = \frac{M}{4} + \frac{3}{4}m < M. \tag{10.20}$$

This result tells us that the new function $h(x, t)$ also attains its maximum value at some point (x_1, t_1) with $0 < x_1 < L, 0 < t_1 \leq T$. At this point we must have that $h_t \geq 0$ and $h_{xx} \leq 0$. [Important question: why don't we just say $h_t = 0$?] Combining these two conditions, we obtain

$$h_t - h_{xx} \geq 0. \tag{10.21}$$

On the other hand, recalling the definition (10.19), we have

$$h_t - h_{xx} = g_t - g_{xx} - \frac{M - m}{2L^2} = -\frac{M - m}{2L^2} < 0. \tag{10.22}$$

Thus, the assumption $M > m$ has led us to a contradiction and the first part of the theorem has been proved. The proof of the part dealing with the minimum value follows directly by noting that if g is a solution so is $-g$. □

Corollary 10.1 (Uniqueness) *The initial-boundary value problem of the heat equation has a unique solution.*

Proof The problem we are describing corresponds to the specification of the values of g at the base and the vertical sides of the rectangle. The proof follows immediately from the assumption that there exist two different solutions to this problem. The difference between these two solutions would, therefore, vanish on this part of the boundary. Since our equation is linear, this difference satisfies the heat equation in the given domain. It follows that (unless the difference is identically zero) we have found a solution of the heat equation that attains a maximum or minimum value at a point not belonging to the part of the boundary stipulated by the maximum–minimum theorem. □

From both the physical and the computational points of view, it is important to ascertain that the behaviour of the solutions of the heat equation is not chaotic. In other words, a small change in the initial and/or boundary data results in a correspondingly small change in the solution. This is the content of the following corollary of the maximum–minimum theorem.

Corollary 10.2 (Continuous dependence on the data) *The solution of the initial-boundary value problem depends continuously on the data in the sense that a small change in the data results in a correspondingly small change in the solution.*

Proof More precisely, this corollary states that if the data (specified on the base and the two vertical sides of the rectangular region $[0, L] \times [0, T]$) corresponding to two solutions $g_1(x, t)$ and $g_2(x, t)$ satisfy the conditions

$$|g_1(0, t) - g_2(0, t)| < \epsilon, \tag{10.23}$$

$$|g_1(x, 0) - g_2(x, 0)| < \epsilon, \tag{10.24}$$

and

$$|g_1(L, t) - g_2(L, t)| < \epsilon, \tag{10.25}$$

then so do the solutions over the whole rectangular domain, viz.:

$$|g_1(x, t) - g_2(x, t)| < \epsilon, \qquad \forall (x, t) \in \mathcal{D}. \tag{10.26}$$

To prove this corollary we start by noticing once again that the difference between two solutions is itself a solution of the (homogeneous) heat equation. The corresponding data are, clearly, given by the difference of the data of the individual solutions. Applying the main theorem to the difference between the given solutions the corollary follows. □

Remark 10.1 A cursory reading of the proof of the maximum–minimum theorem may convey the impression that its main argument could have been applied also to the time-reversed problem. Could one not, having specified the data at the *top* of the rectangle (rather than at its base) and at the vertical sides, conclude that the solution must attain its maximum and minimum values there? The argument appears to be entirely reversible. Nevertheless, when establishing the conditions for a maximum to possibly exist at the base (which step would be necessary in the proof of the theorem), we would have to state that $h_t \leq 0$ and $h_{xx} \leq 0$, thus ruining the conclusion of the theorem. Once again, this is a manifestation of the time-irreversibility of the heat equation.

We have extracted several important facts out of a relatively simple proof. Moreover, the statement of the main theorem corresponds to the physically intuitive fact that if, for example, the maximum value of the temperature data occurs at the base of the rectangle, then we do not expect at any time and at any point the temperature to rise above this value. In its search for thermal equilibrium, the bar will seek to even out the temperatures as much as permitted by the boundary data.

10.5 The Finite Rod

Consider a rod of finite length occupying the interval $[0, L]$ and thermally insulated on its lateral surface. The temperature distribution $g(x, t)$ abides by the diffusion equation (10.4), also called the heat equation. At the ends of the bar, the temperature is kept equal to zero[7] at all times, i.e.,

$$g(0, t) = g(L, t) = 0, \qquad \forall t > 0. \tag{10.27}$$

Moreover, at the initial time, the temperature throughout the length of the rod is given as some continuous function

$$g(x, 0) = g_0(x) \qquad 0 \leq x \leq L. \tag{10.28}$$

For consistency, we assume that the function g_0 vanishes at the two ends of the rod. As we already know (from the results of the previous section), if this problem has a solution it must be unique. We try a solution by the method of separation of variables. According to Eq. (10.17), except for the spatially linear solution, any variable-separated solution must be of the form

$$g(x, t) = (A\cos(\lambda x) + B\sin(\lambda x))\ e^{-\lambda^2 t}. \tag{10.29}$$

Enforcing the boundary conditions (10.27), we obtain that the constant A must vanish and that the parameter λ must belong to a discrete spectrum given by the formula

$$\lambda = \frac{n\pi}{L} \qquad n = 1, 2, \ldots \tag{10.30}$$

[7]Note that the temperature appearing in the heat equation is not necessarily the absolute thermodynamic temperature.

To satisfy the initial condition, we propose an infinite superposition of variable-separated solutions, namely,

$$g(x, t) = \sum_{n=1}^{\infty} B_n \sin\left(\frac{n\pi x}{L}\right) e^{\left(\frac{n\pi}{L}\right)^2 t}. \tag{10.31}$$

Introducing this form of the solution into the initial condition (10.28), we obtain

$$g_0(x) = \sum_{n=1}^{\infty} B_n \sin\left(\frac{n\pi x}{L}\right). \tag{10.32}$$

By the orthogonality property of the trigonometric functions involved, we have

$$B_n = \frac{2}{L} \int_0^L g_0(x) \sin\left(\frac{n\pi x}{L}\right) dx. \tag{10.33}$$

Recall that we have simplified the heat equation by assuming that $D = 1$. If we now restore the original value of this constant, the only change to Eq. (10.31) would consist of a rescaling of the time variable, that is,

$$g(x, t) = \sum_{n=1}^{\infty} B_n \sin\left(\frac{n\pi x}{L}\right) e^{-\left(\frac{n\pi}{L}\right)^2 Dt}. \tag{10.34}$$

Notice that as time goes on, the solution tends to a state of thermal equilibrium, as expected. From a detailed analysis of this solution, one can verify that if the initial temperature distribution is continuous with piece-wise continuous derivatives, the solution is of class C^{∞} for $t > 0$. We have already alluded to this property earlier and indicated that, from the physical point of view, its meaning is that any irregularities in the initial data are immediately smoothed out by the diffusive process of heat transfer. It is interesting to remark that one can use this property of the solution to prove that the heat equation cannot in general be solved backward in time.

The method of separation of variables has allowed us to solve the homogeneous heat equation (no distributed heat sources or sinks) under a regime of homogeneous boundary conditions (zero temperature at the ends of the rod). Other, more general, homogeneous boundary conditions (such as insulated ends) can also be considered, leading to Fourier series expansions involving cosine terms. The solution of cases where the boundary conditions are arbitrary functions of time or where there exist heat sources distributed over the length of the rod, however, requires the consideration of other methods (such as Laplace transforms, Green functions, Duhamel integrals and eigenfunction expansions). The general treatment of these methods is beyond the scope of this book, but we will explore some of them to a limited extent.

10.6 Non-homogeneous Problems

Although more general non-homogeneous boundary conditions can be considered, we will limit our attention to boundary conditions of temperature. We consider, therefore, the problem

$$g_t - Dg_{xx} = 0 \qquad 0 < x < L \qquad t > 0, \tag{10.35}$$

$$g(x, 0) = g_0(x) \qquad 0 \leq x \leq L, \tag{10.36}$$

$$g(0, t) = f_0(t) \qquad g(L, t) = f_L(t) \qquad t \geq 0. \tag{10.37}$$

We assume, moreover, that the following consistency conditions are satisfied, namely,

$$g_0(0) = f_0(0) \qquad g_0(L) = f_L(0). \tag{10.38}$$

Our aim is to show that this problem (a homogeneous equation with inhomogeneous boundary conditions) can be generally converted into a problem of a non-homogeneous equation with homogeneous boundary conditions. To this effect, we decompose the solution into the sum of two terms as

$$g(x, t) = G(x, t) + S(x, t). \tag{10.39}$$

The second term (which, somewhat imprecisely, will be referred to as the steady state part of the solution) is given by a spatial linear interpolation of the given boundary conditions, namely,

$$S(x, t) = f_0(t)\left(1 - \frac{x}{L}\right) + f_L(t)\,\frac{x}{L}. \tag{10.40}$$

Introducing the proposed decomposition (10.39) into the original PDE (10.35), we obtain

$$G_t - DG_{xx} = -S_t. \tag{10.41}$$

Thus, the "transient" (non-steady) part, $G(x, t)$, of the solution satisfies a non-homogeneous version of the heat equation, whereby the sources are obtained by a particular linear combination of the time-derivatives of the boundary conditions. On the other hand, it is not difficult to verify that (in fact, by construction) the function $G(x, t)$ satisfies the homogeneous boundary conditions

$$G(0, t) = G(L, t) = 0 \qquad \forall t > 0, \tag{10.42}$$

and the initial condition

$$G(x, 0) = g_0(x) - S(x, 0) \qquad 0 \leq x \leq L. \tag{10.43}$$

We conclude that a heat conduction problem with non-homogeneous boundary conditions can be transformed into a counterpart with homogeneous boundary conditions at the price of introducing distributed and time-dependent heat sources and modifying the initial conditions, both in an elementary manner. The solution of the transformed problem can be achieved by the method of eigenfunction expansion (which, as in the case of the wave equation, we also call normal-mode superposition). The eigenfunctions are precisely the spatial parts of the variable-separated solutions of the homogeneous equation that we studied earlier. In the case of a bar with uniform properties, these eigenfunctions are harmonic functions, thus leading to the Fourier series expansion. In more general cases (which we will not consider) the eigenfunctions are not harmonic, but (according to the Sturm–Liouville theory) still constitute a complete set of orthogonal functions.

In order to solve the problem given by Eqs. (10.41)–(10.43), we express a particular solution as

$$G(x, t) = \sum_{n=1}^{\infty} D_n(t) \sin\left(\frac{n\pi x}{L}\right). \tag{10.44}$$

This expression agrees with its counterpart for the treatment of the non-homogeneous wave equation that we studied in a previous chapter. Similarly, we express the heat source as

$$S_t(x, t) = \sum_{n=1}^{\infty} C_n(t) \sin\left(\frac{n\pi x}{L}\right). \tag{10.45}$$

The coefficients of this expansion can be calculated at each instant of time by the by now familiar formula

$$C_n(t) = \frac{2}{L}\int_0^L S_t(x, t) \sin\left(\frac{n\pi x}{L}\right) dx. \tag{10.46}$$

Introducing the representations (10.45) and (10.46) into (10.41), we obtain a sequence of mutually independent ODEs, that is,

$$\frac{dD_n}{dt} + D\left(\frac{n\pi}{L}\right)^2 D_n = -C_n. \tag{10.47}$$

A particular solution of this equation is given by

$$D_n(t) = -\int_0^t C_n(\tau) e^{-\left(\frac{n\pi}{L}\right)^2 D(t-\tau)}\, d\tau. \tag{10.48}$$

The complete solution of the non-homogeneous problem is, therefore,

$$G(x,t) = \sum_{n=1}^{\infty} \left(B_n e^{-\left(\frac{n\pi}{L}\right)^2 Dt} + D_n(t) \right) \sin\left(\frac{n\pi x}{L}\right). \tag{10.49}$$

The constants can be adjusted to satisfy the initial condition (10.43). Finally, the solution of the original problem is obtained form Eq. (10.39).

10.7 The Infinite Rod

In the case of a rod of infinite spatial extent, we are confronted with the pure Cauchy (or initial-value) problem

$$g_t - Dg_{xx} = 0 \qquad -\infty < x < \infty \qquad t > 0, \tag{10.50}$$

$$g(x,0) = g_0(x) \qquad -\infty < x < \infty, \tag{10.51}$$

without any boundary conditions. We will assume that the initial temperature distribution $g_0(x)$ is continuous and bounded over the real line. To show the uniqueness of the solution of this problem, we would like to emulate the procedure we used in the case of the finite rod, namely, to prove a maximum–minimum theorem. In order to achieve this goal, however, it turns out that, unlike the finite case, we must now make an extra assumption on the nature of the solution: we need to assume a-priori that the sought after solution is continuous and bounded. Otherwise, it can be shown explicitly that the maximum–minimum theorem doesn't hold and the solution is, in fact, not unique. A standard argument due to Tychonoff[8] shows how to construct a C^∞ solution that vanishes at $t = 0$. This solution, however, is unbounded. A solution $g(x,t)$ is said to be bounded if there exists a positive number M such that

$$|g(x,t)| \le M \qquad -\infty < x < \infty \qquad t > 0. \tag{10.52}$$

Let $g_1(x,t)$ and $g_2(x,t)$ be two bounded solutions of Eq. (10.50). The difference $g = g_1 - g_2$ between these solutions is, therefore, also bounded. Instead of proceeding to prove an independent maximum theorem, we can take advantage of the maximum theorem for the finite case to produce a proof of uniqueness by showing that $g(x,t)$ must vanish identically over the half plane of interest. To this end, we attempt to construct a family of solutions $g_L(x,t)$ of the heat equation, over the finite spatial intervals $-L \le x \le L$, each of which enjoys the property of being non-negative

[8] See [4], p. 211.

and greater than (or equal to) $|g(x,t)|$ over the common domain of definition. Such a family of solutions is given by the prescription[9]

$$g_L(x,t) = \frac{4M}{L^2}\left(Dt + \frac{x^2}{2}\right), \tag{10.53}$$

as can be verified. In particular, we notice that for any given $T > 0$ the values taken by this solution over the part of the boundary consisting of the base $[-L, L] \times \{0\}$ and the sides $\{-L\} \times [0, T]$ and $\{L\} \times [0, T]$ are point by point larger than (or equal to) the corresponding values of $|g|$. This must, therefore, be true for the interior points as well. Fixing an arbitrary point (x, t), we conclude from Eq. (10.53) that for sufficiently large L the absolute value of $g(x, t)$ can be bounded by as small a positive number as desired. This concludes the proof of uniqueness. As a corollary of this theorem, one can (by the same procedure as in the finite case) prove the continuous dependence of the solution on the initial data.

Having thus demonstrated the uniqueness of the Cauchy problem, we need to construct a solution by any method, which will then become the (unique) solution. In particular, if two solutions are found which appear to be different (perhaps because of the different methods used to derive them), they are automatically identical to each other. We have already remarked that fairly general solutions of the heat equation can be found by adjusting the coefficients in the expression

$$g(x,t) = \int_{-\infty}^{\infty} \left(A(\lambda)\cos(\lambda x) + B(\lambda)\sin(\lambda x)\right) e^{-\lambda^2 t}\, d\lambda, \tag{10.54}$$

so as to match the initial condition $g(x, 0) = g_0(x)$, namely,

$$g_0(x) = \int_{-\infty}^{\infty} \left(A(\lambda)\cos(\lambda x) + B(\lambda)\sin(\lambda x)\right)\, d\lambda. \tag{10.55}$$

When we compare this expression with the familiar formula for the Fourier series, we realize that it can be regarded as a generalized version of it. The generalization consists in not demanding that the function represented be periodic, since the domain of definition of the initial conditions is now unbounded. As a result, we no longer obtain a discrete spectrum of possible values for the wavelength, but rather a continuous spectrum, where every wavelength is represented. If we had at our disposal some kind of orthogonality condition, as was the case in the finite domain, we would be able to obtain these coefficients directly from Eq. (10.55). Instead, we will proceed to introduce the concept of *Fourier integral* by a heuristic argument of passage to the limit of the Fourier series as the period tends to infinity.

---

[9]As suggested in [6], p. 605.

10.8 The Fourier Series and the Fourier Integral

A periodic function $f(x)$ of period $2L$ can be represented by means of a Fourier series in the form

$$f(x) = \frac{1}{2}a_0 + \sum_{n=1}^{\infty} (a_n \cos(\lambda_n x) + b_n \sin(\lambda_n x)) . \tag{10.56}$$

The equal sign in this equation has to be taken with a pinch of salt. Be that as it may, the "frequencies" λ_n constitute a discrete spectrum dictated by the period of the function being represented, specifically given by

$$\lambda_n = \frac{n\pi}{L}. \tag{10.57}$$

The coefficients of the expansion (10.57), also called amplitudes, are given by the integrals

$$a_n = \frac{1}{L}\int_{-L}^{L} f(\xi)\cos(\lambda_n \xi)\, d\xi, \tag{10.58}$$

$$b_n = \frac{1}{L}\int_{-L}^{L} f(\xi)\sin(\lambda_n \xi)\, d\xi. \tag{10.59}$$

These formulas are obtained by a direct application of the orthogonality property of trigonometric functions and by assuming that the Fourier series can be integrated term by term. It is convenient to write the Fourier series (10.56) in a more compact notation, using complex algebra. Recalling the identity

$$e^{i\alpha} = \cos\alpha + i\sin\alpha, \tag{10.60}$$

(where i denotes the imaginary unit), we can write (10.56) as

$$f(x) = \sum_{n=-\infty}^{\infty} c_n e^{i\lambda_n x}. \tag{10.61}$$

The (complex) coefficients are related to the (real) coefficients by the formulas

$$c_n = \begin{cases} \frac{1}{2}(a_n - ib_n) \text{ for } n \geq 0 \\ \frac{1}{2}(a_n + ib_n) \text{ for } n < 0 \end{cases} \tag{10.62}$$

More explicitly, the coefficients are given by

$$c_n = \frac{1}{2L} \int_{-L}^{L} f(\xi) e^{-i\lambda_n \xi} \, d\xi. \tag{10.63}$$

Notice that, although the original function may have been defined only in the interval $[-L, L]$, the Fourier representation is valid over the whole line. In other words, the Fourier series represents a periodic extension of the given function, obtained by just translating and copying the function ad infinitum. When performing this extension, even if the original function is continuous, we may obtain points of discontinuity at the extreme values of each period. In such cases, it can be shown that the Fourier series converges to the average value. We will not discuss these or other phenomena pertaining to the convergence of Fourier series. In particular, we will assume that differentiation can be carried out term by term and that the series thus obtained is an almost-everywhere faithful representation of the derivative of the original function. Taking these liberties, we can easily understand why the Fourier series can be so useful in the solution of differential equations. For example, the second derivative of a function has Fourier coefficients which are, one by one, proportional to the coefficients of the original function, the constant of proportionality being $-\lambda_n^2$. In the transformed world of Fourier coefficients, therefore, taking a second derivative is interpreted as a kind of multiplication.

We want to extend the above concepts to functions that are not necessarily periodic and that are defined over the entire real line. We will make the assumption that the absolute value of the given function $f(x)$ is integrable and that the integral over the real line is finite, namely, for some positive number M,

$$\int_{-\infty}^{\infty} |f(x)| \, dx \leq M. \tag{10.64}$$

Let us consider an arbitrary interval $[-H, H]$. If we restrict the given function to this interval and then extend it periodically, we can represent the resulting function $f_H(x)$ by means of a Fourier series as

$$f_H(x) = \frac{1}{2H} \sum_{n=-\infty}^{\infty} \int_{-H}^{H} f(\xi) e^{-\frac{in\pi(\xi - x)}{H}} \, d\xi, \tag{10.65}$$

where we have combined Eqs. (10.61) and (10.62). We intend to let H go to infinity and to replace the summation by an integral. To achieve this goal, it is convenient to define

$$\Delta = \frac{\pi}{H}. \tag{10.66}$$

We rewrite Eq. (10.65) trivially as

$$f_H(x) = \frac{1}{2\pi} \sum_{-\infty}^{\infty} \left(\Delta \int_{-H}^{H} f(\xi) e^{\frac{in\pi(x-\xi)}{H}} \, d\xi \right). \tag{10.67}$$

But if we recall the definition of the (Riemann) integral of a function as a limit of sums, we obtain (as $\Delta \to 0$ while $H \to \infty$)

$$f(x) = \frac{1}{2\pi} \int_{-\infty}^{\infty} \int_{-\infty}^{\infty} f(\xi) e^{i\lambda(x-\xi)} \, d\xi \, d\lambda. \tag{10.68}$$

This formula is known as the Fourier integral. Notice that in obtaining this result we have defined a new continuous variable λ, whose discrete values in the limiting process were $n\pi/H = n\Delta$, precisely as required by the definition of an integral. The Fourier integral can be regarded in two steps, just as we suggested for the Fourier series. In the first step, called the *Fourier transform*, we produce a transformation of the original function f (of the independent variable x) to another function F (of the independent variable λ running within the "frequency domain") by means of the formula

$$F(\lambda) = \frac{1}{\sqrt{2\pi}} \int_{-\infty}^{\infty} f(\xi) e^{-i\lambda\xi} \, d\xi. \tag{10.69}$$

The second step consists of the *inverse Fourier transform*

$$f(\xi) = \frac{1}{\sqrt{2\pi}} \int_{-\infty}^{\infty} F(\lambda) e^{i\lambda\xi} \, d\lambda. \tag{10.70}$$

The process of obtaining the Fourier transform of functions is, clearly, a linear operation from the space of functions into itself. We indicate this linear operator by $\mathcal{F}$. Thus, we can write

$$F(\lambda) = \mathcal{F}[f(x)]. \tag{10.71}$$

Given a differential equation, we can apply the Fourier transform to hopefully obtain a tractable problem in the frequency domain. After solving this problem, we may attempt to return to the real world by means of the inverse transform. Before entering any such considerations, we should remember that the Fourier integral was obtained by a heuristic process of passage to the limit, so that it behooves us to justify this process by checking directly that the result is correct. We will, however, omit this (not very difficult) proof.[10]

---

[10]See [1], p. 78.

Consider the Fourier transform of the derivative of a function. We have:

$$\mathcal{F}[f'(x)] = \frac{1}{\sqrt{2\pi}} \int_{-\infty}^{\infty} f'(\xi)e^{-i\lambda\xi}\, d\xi = \frac{1}{\sqrt{2\pi}} \left(f(\xi)e^{-i\lambda\xi}\Big|_{-\infty}^{\infty} + i\lambda \int_{-\infty}^{\infty} f(\xi)e^{-i\lambda\xi}\, d\xi \right). \tag{10.72}$$

If the original function vanishes at $\pm\infty$, we obtain the important relation

$$\mathcal{F}[f'(x)] = i\lambda\mathcal{F}[f(x)]. \tag{10.73}$$

Again, just as in the case of the Fourier series, we obtain that in the frequency domain differentiation is interpreted as multiplication by the frequency variable. Another important property is the so-called *convolution*. The convolution product $f * g$ of two functions $f(x)$ and $g(x)$ is defined as

$$(f * g)(x) = \frac{1}{\sqrt{2\pi}} \int_{-\infty}^{\infty} f(x-\xi)g(\xi)\, d\xi. \tag{10.74}$$

The convolution product is commutative and associative. The Fourier transform of the convolution product of two functions can be shown to be equal to the product of the transforms, that is,

$$\mathcal{F}[f * g] = \mathcal{F}[f]\, \mathcal{F}[g]. \tag{10.75}$$

10.9 Solution of the Cauchy Problem

We are now in a position of solving the Cauchy problem as formulated in Eqs. (10.50) and (10.51), provided we add the extra condition that both the initial temperature distribution $g_0(x)$ and its derivative $g_0'(x)$ vanish at $x = \pm\infty$. Fourier-transforming Eq. (10.50) with respect to the space variable (while the time variable remains as a parameter), we can write

$$\mathcal{F}[g_t(x,t)] - D\mathcal{F}[g_{xx}(x,t)] = 0. \tag{10.76}$$

We note that the x-Fourier transform of $g(x,t)$ is a function of λ and of the time variable t, which has remained unaffected by the transformation. Denoting this transform by $G(\lambda,t)$ and using the derivative property of the Fourier transform, we write

$$G_t(\lambda,t) + D\lambda^2 G(\lambda,t) = 0. \tag{10.77}$$

We note that the derivative with respect to the time parameter is directly reflected as the derivative with respect to the same parameter of the Fourier transform, and it is only the derivative with respect to the transformed variable that enjoys the special property derived above. For each value of λ, Eq. (10.77) is a first-order ODE. The initial condition is obtained by Fourier-transforming the initial condition (10.51). We denote

$$G_0(\lambda) = \mathcal{F}[g_0(x)]. \tag{10.78}$$

The solution of (10.77) with initial condition (10.78) is an elementary problem and we obtain

$$G(\lambda, t) = G_0(\lambda)e^{-D\lambda^2 t}. \tag{10.79}$$

The solution to the original problem is given by the inverse transform of this function. The calculation of inverse transforms is usually not a straightforward task. In this particular case, however, it can be accomplished by an application of the convolution formula (10.75). Indeed, the inverse transform of the second factor in the right-hand side of (10.79) is given by

$$\mathcal{F}^{-1}\left[e^{-D\lambda^2 t}\right] = \frac{1}{\sqrt{2Dt}}\, e^{-\frac{x^2}{4Dt}}. \tag{10.80}$$

We have obtained this value from a table of Fourier transforms, although in this case the direct evaluation of the inverse transform is relatively straightforward. Applying now the convolution formula to (10.79), we obtain

$$g(x,t) = \mathcal{F}^{-1}\left[G_0(\lambda)e^{-D\lambda^2 t}\right] = g_0(x) * \frac{1}{\sqrt{sDt}}\, e^{-\frac{x^2}{4Dt}}, \tag{10.81}$$

```
Plot3D[(1/(2*Sqrt[Pi*t]))* NIntegrate[Exp[(-x^2/2 - (x - ksi)^2/(4*t))], {ksi, -3, 3}],
{x, -2.5, 2.5}, {t, 0.0001, 50}, AxesLabel -> {x, t, u}, PlotRange -> All]
```

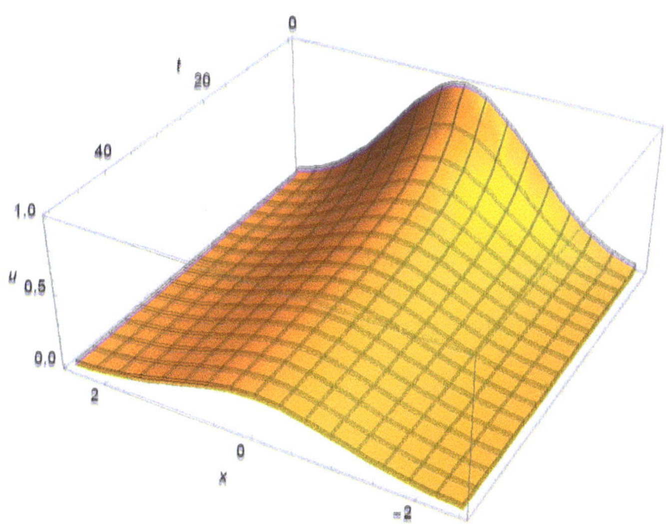

Fig. 10.3 Plot of Eq. (10.82)

or

$$g(x,t) = \frac{1}{2\sqrt{\pi D t}} \int_{-\infty}^{\infty} g_0(\xi) e^{-\frac{(x-\xi)^2}{4Dt}} \, d\xi. \tag{10.82}$$

This is the solution of the Cauchy problem for the heat equation. A useful way to interpret this result is obtained by the use of the concept of generalized functions, as we will do in the next section. Figure 10.3 shows a plot of the solution (10.82) for $D = 1$ and a bell-shaped initial temperature distribution given by the function $g_0(x) = e^{-x^2/2}$. The integration has been numerically achieved with the use of Mathematica®. The initial time has been taken somewhat greater than zero, to avoid a numerical singularity.

10.10 Generalized Functions

We define a *generalized function* (or a *distribution*) as a real-valued linear functional on a space F of functions.[11] One of the physical motivations behind this notion is the ability to describe concentrated entities (forces, masses, and so on) within the same framework as smoothly distributed ones. Let us denote by ϕ one such generalized function. By saying that it is a real-valued functional on a space of functions we mean that, given a function f belonging to this space, the functional ϕ assigns to it a real number, which we will denote as $\phi[f]$. The fact that this functional is linear means that, given any two functions, f and g, and any two real numbers α and β, we must have

$$\phi[\alpha f + \beta g] = \alpha\phi[f] + \beta\phi[g]. \tag{10.83}$$

Given a continuous (or just integrable) function f, we can assign to it a unique generalized function ϕ_f by means of the prescription

$$\phi_f = \int_{-\infty}^{\infty} g(x) f(x) \, dx \qquad \forall g \in F. \tag{10.84}$$

It is not difficult to prove (by an argument akin to the so-called fundamental lemma of the calculus of variations) that the linear functional thus defined determines the function f uniquely. If this were all that we have to say about generalized functions it wouldn't be worth our effort. But consider, for example, the following functional on the given space of functions F: it assigns to each function the value of the function at the origin. We denote this functional by δ and call it *Dirac's delta*. More precisely

[11] For technical reasons, the space of functions over which these functionals are defined consists of the so-called *space of test functions*. Each test function is of class C^∞ and has *compact support* (that is, it vanishes outside a closed and bounded subset of $\mathbb{R}$). The graph of a test function can be described as a smooth 'bump'.

$$\delta[g] = g(0) \qquad \forall g \in F. \tag{10.85}$$

This functional is clearly linear, but is not amenable to an integral representation in the conventional sense. In other words, there is no ordinary function, f_δ say, which, when plugged into the integrand of Eq. (10.84) will produce the same result as Dirac's delta. Nevertheless, we can now symbolically imagine a "function" $\delta(x)$ which somehow does the trick, namely,

$$\int_{\infty}^{\infty} g(x)\delta(x)\,dx = g(0) \qquad \forall g \in F. \tag{10.86}$$

A useful way to look at this integral is to consider Dirac's delta "function" as a limit of a sequence of integrable functions $\delta_n(x)$ $(n = 1, 2, \ldots)$ that vanish outside increasingly smaller intervals around the origin. This is, after all, the "physical" meaning of a concentrated entity. Such a sequence of functions, illustrated in Fig. 10.4, can be defined as

$$\delta_n(x) = \begin{cases} \frac{n}{2} \text{ if } |x| < \frac{1}{n} \\ 0 \text{ if } |x| \geq \frac{1}{n} \end{cases} \tag{10.87}$$

The area under the graph of each function remains thus always equal to 1. As we calculate the integral of the product of these functions with any given function g, we obtain

$$\int_{-\infty}^{\infty} g(x)\delta_n(x)\,dx = \frac{n}{2}\int_{-1/n}^{1/n} g(x)\,dx = g(\bar{x}). \tag{10.88}$$

We have used the mean value theorem to replace the integral of a function by the value of the function at an interior point $\bar{x}$ times the length $2/n$ of the interval of integration. As n increases, the intermediate point gets more and more confined and eventually becomes the origin, which is the only point common to all the nested intervals. Thus we recover (10.86) as a limiting case.

An obvious feature of the Dirac delta function is its *filtering* or substitution property

$$\int_{-\infty}^{\infty} g(x)\delta(x-a)\,dx = g(a). \tag{10.89}$$

Generalized functions are in general not differentiable, but we are interested in extending the notion of differentiability in such a way that derivatives may be defined in a distributional sense. To obtain a meaningful definition, we can emulate the properties of the derivatives of those distributions which do have ordinary derivatives, namely, those given by Eq. (10.84). For these regular distributions, we clearly want to have the property

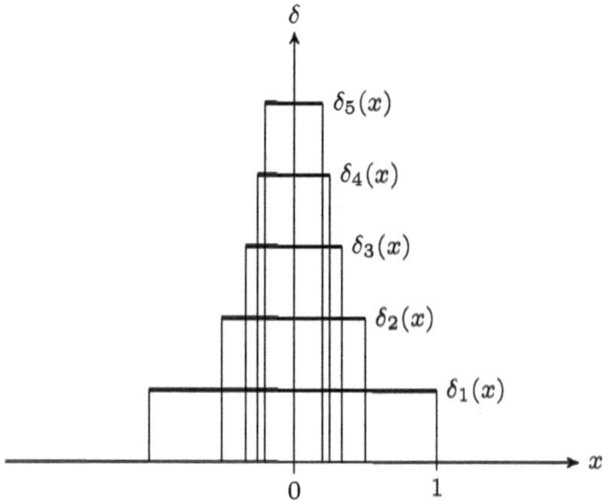

Fig. 10.4 Intuiting the Dirac distribution

$$\phi_{f'}[g] = \int_{-\infty}^{\infty} g(x) f'(x)\, dx \qquad \forall g \in F. \tag{10.90}$$

At this point, we can integrate by parts[12] and obtain

$$\phi_{f'}[g] = -\int_{-\infty}^{\infty} g'(x) f(x)\, dx = -\phi_f[g'] \qquad \forall g \in F. \tag{10.91}$$

Accordingly, we adopt the following definition for the derivative of a distribution

$$\phi'[g] = -\phi[g'] \qquad \forall g \in F. \tag{10.92}$$

The distributional derivative of a distribution is itself a distribution. An important example is given by the distributional derivative of the Heaviside (step) function

$$H(x) = \begin{cases} 0 \text{ if } x < 0 \\ 1 \text{ if } x \geq 0 \end{cases} \tag{10.93}$$

The generalized derivative (which we don't bother to indicate with anything but the conventional symbol for ordinary derivatives) is given by[13]

[12] We must now use the fact that the function space consisted of functions with compact support, so that they, and all their derivatives, vanish at infinity.

[13] Here again we are using the compact support property.

$$H'[g] = -H[g'] = -\int_{-\infty}^{\infty} g'(x)H(x)\,dx = -\int_{0}^{\infty} g'(x)\,dx = g(0). \tag{10.94}$$

We see that the action of the derivative of the Heaviside function is identical to the action of the Dirac delta function on each and every function of the original function space. We conclude, therefore, that the derivative of the Heaviside function is the Dirac delta function.

Let us go back to our solution of the heat equation as expressed in Eq. (10.82), and let us assume that our initial condition $g_0(x)$ is not a function but a distribution. In particular, let us consider the case

$$g_0(x) = \delta(x-a). \tag{10.95}$$

The physical meaning of such an initial condition is that at the initial time we placed a concentrated source of heat at the point $x = a$. This interpretation is clearly contained in the conception of the Dirac function as a limit of a sequence of ordinary functions, as we have demonstrated above. Indeed, the functions in this sequence vanish everywhere except for an increasingly smaller and smaller interval around that point. When we plug this initial condition in the general solution of the Cauchy problem, we obtain:

$$g_a(x,t) = \frac{1}{2\sqrt{\pi D t}} \int_{-\infty}^{\infty} \delta(\xi - a) e^{-\frac{(x-\xi)^2}{4Dt}}\,dx = \frac{e^{-\frac{(x-a)^2}{4Dt}}}{2\sqrt{\pi D t}}. \tag{10.96}$$

The meaning of the expression in the right-hand side of this equation is, therefore, the temperature distribution, as a function of time and space, in an infinite rod which has been subjected to a concentrated unit source of heat at the point $x = a$ at time $t = 0$. This is thus some kind of "influence function" (of the same type that used to be studied in structural engineering in the old days for bridge design). In the context of the theory of differential equations, these functions (representing the effect due to a unit concentrated cause at an arbitrary position) are called Green's functions. The usefulness of Green's functions is that, because the differential equation of departure is linear, we can conceive of the solution as simply a superposition of the effects of the concentrated unit sources. This interpretation is borne out by the following equation

$$g(x,t) = \int_{-\infty}^{\infty} g_0(a) g_a(x,t)\,da, \tag{10.97}$$

which is the same as Eq. (10.82). This calculation shows that, if we have any means (exact or approximate) to calculate Green's function for a particular differential equation (perhaps with some boundary conditions), then we have a recipe for constructing solutions by superposition integrals.

10.11 Inhomogeneous Problems and Duhamel's Principle

Consider once more the Cauchy problem[14] for the inhomogeneous heat equation

$$g_t - Dg_{xx} = f(x,t) \qquad -\infty < x < \infty \qquad t > 0, \tag{10.98}$$

with the initial conditions

$$g(x,0) = g_0(x) \qquad -\infty < x < \infty. \tag{10.99}$$

As we have already discovered in Sect. 8.9 when dealing with the wave equation, Duhamel's principle constructs the solution of non-homogeneous problems out of a clever continuous superposition of solutions of homogeneous problems, for which the solution is assumed to be known (either exactly or approximately). We remark that we only need to solve the stated problem (10.98) for homogeneous initial conditions, i.e., for

$$g(x,0) = 0 \qquad -\infty < x < \infty. \tag{10.100}$$

Indeed, if we consider first the homogeneous problem

$$g_t - Dg_{xx} = 0 \qquad -\infty < x < \infty \qquad t > 0, \tag{10.101}$$

with the original initial conditions (10.99) as solved, and if we call its solution $\bar{g}(x,t)$, then the function $g(x,t) - \bar{g}(x,t)$ satisfies Eq. (10.98) with the homogeneous initial conditions (10.100). In other words, if we solve Eq. (10.98) with homogeneous initial conditions, all we have to do to obtain the required solution is to add the solution of the homogeneous equation with the original inhomogeneous initial conditions.

In view of our newly acquired familiarity with the Dirac distribution, we may motivate Duhamel's principle by viewing the right-hand side of Eq. (10.98), representing the forcing function, as an infinite superposition of pulses in the form

$$f(x,t) = \int_{-\infty}^{\infty} f(x,\tau)\delta(t-\tau)\, d\tau. \tag{10.102}$$

For this integral to make sense, we are tacitly extending the given forcing function as zero over the interval $(-\infty, 0)$. At any rate, if $t > 0$, the lower limit of the integral can be changed to 0. The mathematical expression (10.102) can be seen as the counterpart of the graphical representation given in Box 8.2.

Assume that we are able to solve the inhomogeneous problem

$$g_t - Dg_{xx} = f(x,\tau)\delta(t-\tau) \qquad -\infty < x < \infty \qquad t > \tau > 0, \tag{10.103}$$

[14] Although we are presenting the principle in the context of an infinite rod, the same idea can be applied to the case of the finite rod.

with vanishing initial conditions at $t = \tau$. We denote the solution of this problem, in which τ acts just as a parameter, $g(x, t; \tau)$. Notice that this function vanishes identically for $t < \tau$. Then, by superposition, we must have

$$g(x,t) = \int_0^t g(x,t;\tau)\,d\tau. \tag{10.104}$$

The remarkable fact is that we can actually obtain the solution of (10.103) by means of an initial value problem of the homogeneous equation! To visualize how this is possible,[15] all we need to do is integrate Eq. (10.103) with respect to time over a small interval $(\tau - \epsilon, \tau + \epsilon)$. In so doing, and taking into consideration that $g(x, t; \tau)$ vanishes for $t < \tau$, we obtain

$$g(x,\tau+\epsilon;\tau) - 2\epsilon D\bar{g}_{xx} = f(x,\tau), \tag{10.105}$$

where $\bar{g}_{xx}$ is an intermediate value within the given interval. As $\epsilon \to 0$, we conclude that

$$g(x,\tau;\tau) = f(x,\tau). \tag{10.106}$$

The meaning of this equation is that the problem of Eq. (10.103) with homogeneous initial conditions can be replaced with the homogeneous problem (no forcing term), but with the initial condition

$$g(x,0;\tau) = f(x,\tau). \tag{10.107}$$

Since we have assumed that the solution of the homogeneous problem with arbitrary initial conditions is available (for example, by means of Fourier transforms, as per Eq. (10.82)), we obtain the solution of the original problem by means of Eq. (10.94).

Exercises

Exercise 10.1 (*Spinal drug delivery*)[16] A drug has been injected into the spine so that at time $t = 0$ it is distributed according to the formula $g(x, 0) = c + C \sin \frac{\pi x}{L}$, where the spine segment under study extends between $x = 0$ and $x = L$, and where c and C are constants. The drug concentration at the ends of the spine segment is artificially maintained at the value c for all subsequent times. If, in the absence of any production, T is the time elapsed until the difference between the concentration at the midpoint of the spine and c reaches one-half of its initial value, calculate the longitudinal diffusion coefficient D of the drug through the spinal meninges. For a rough order of magnitude, assume $L = 10$ mm and $T = 3$ h. [Hint: verify that

[15] This is a somewhat different interpretation from that of Sect. 8.9.

[16] See [2], p. 40.

$g(x, t) = c + C \exp\left(-\frac{D\pi^2 t}{L^2}\right) \sin \frac{\pi x}{L}$ satisfies the PDE (10.4) and the initial and boundary conditions.]

Exercise 10.2 (*Modified discrete model*) Show that if the probability β in the model of Box 10.1 has a value $0 < \beta < 0.5$, so that the particles have a positive probability $\alpha = 1 - 2\beta$ of staying put, the diffusion equation is still recovered, but with a different value for the diffusion coefficient D.

Exercise 10.3 (*Biased discrete diffusion*) Let β^+ and $\beta^- = 1 - \beta^+$ represent, respectively, the generally different probabilities for a particle to move to the right or to the left. Obtain a PDE whose approximation matches the corresponding discrete model. Propose a physical interpretation.

Exercise 10.4 (*Finite domain*) Modify the original discrete model so that it can accommodate a spatial domain of a finite extent. Consider two different kinds of boundary conditions, as follows: (1) The number of particles at each end of the domain remains constant. For this to be the case, new particles will have to be supplied or removed at the ends. (2) The total number of particles is preserved, with no flux of new particles through the end points of the domain. Implement the resulting model in a computer code and observe the time behaviour. What is the limit state of the system for large times under both kinds of boundary conditions?

Exercise 10.5 Prove that if the data (over the three appropriate sides of a rectangular region) of one solution of the heat equation are everywhere greater than the data of another solution, then the same holds true for the corresponding solutions at all the interior points of the region. Moreover, show that if the absolute value of the data at each point is smaller than that of the data of an everywhere positive solution, then so is the absolute value at each interior point smaller than the corresponding value of the (positive) solution at that point.

Exercise 10.6 (*Irreversibility*) Show that the boundary-initial value of the heat equation for a finite rod cannot in general be solved backward in time for general initial conditions. [Hint: assume that it can and impose initial conditions that are not of class C^∞.]

Exercise 10.7 Solve the problem[17]

$$g_t = g_{xx} \qquad 0 < x < 1 \qquad t > 0,$$

$$g(0, t) = 0 \qquad g(1, t) = \cos t \qquad t > 0,$$

$$g(x, 0) = x \qquad 0 \leq x \leq 1.$$

Use the method of eigenfunction expansion.

[17] See [3], p. 71.

Exercise 10.8 Verify the equivalence between expressions (10.56) and (10.61).

Exercise 10.9 (*Convolution*) Prove Eq. (10.75). [Hint: apply the inverse Fourier transform to the right-hand side.]

Exercise 10.10 Solve the Cauchy problem for the non-homogeneous heat equation

$$g_t - Dg_{xx} = f(x,t) \qquad -\infty < x < \infty \qquad t > 0, \tag{10.108}$$

with homogeneous initial conditions by means of Fourier transforms. [Hint: to solve the ODE obtained in the frequency domain, use the method of variation of the constants].

Exercise 10.11 (*Fourier transform and the wave equation*) Apply the Fourier transform method to the solution of the Cauchy problem for the (homogeneous) one-dimensional wave equation when the initial velocity is zero and the initial deformation of the (infinite) string is given as some function $f(x)$. Compare with the d'Alembert solution.

Exercise 10.12 The free transverse vibrations of a beam are described by the fourth-order differential equation (9.80), namely,

$$c^4 u_{xxxx} + u_{tt} = 0,$$

where c is a constant. Use Fourier transforms to solve for the vibrations of an infinite beam knowing that at time $t = 0$ the beam is released from rest with a displacement given by a function $u(x, 0) = f(t)$. Express the result as an integral. Hint: The inverse Fourier transform of $\cos a\lambda^2$ is $\frac{1}{\sqrt{2a}} \cos\left(\frac{x^2}{4a} - \frac{\pi}{4}\right)$.

Exercise 10.13 (*Distributional derivative*) Justify the conclusion that the distributional derivative of the Heaviside function is the Dirac distribution by approximating the (discontinuous) Heaviside function by means of a sequence of continuous functions including straight transitions with increasingly steeper slopes.

Exercise 10.14 (*Bending moment diagrams*) If a bending moment diagram of a beam is a C^2 function, considerations of equilibrium show that its second derivative equals the applied distributed load. Imagine now a C^0 bending moment diagram given by the formula

$$M(x) = \begin{cases} kx & 0 \le x \le L/2 \\ k(L-x) & L/2 < x \le L \end{cases}$$

where k is a constant and L is the length of the beam. Determine the load applied on the beam by calculating the second distributional derivative of the given bending moment function. Show that, extending the diagram as zero beyond the beam domain, we also recover the reactions at the supports.

Exercise 10.15 (*Influence function in statics*) Given a simply supported beam, identify the influence function $g_a(x)$ with the bending moment diagram due to a unit concentrated force acting at the point $a \in [0, L]$. Apply Eq. (10.97) carefully to obtain the bending moment diagram for an arbitrary load $g(x)$. Compare with the standard answer. Check the case $g(x) = \text{constant}$.

Exercise 10.16 Express the solution of the inhomogeneous problem at the end of Sect. 10.9 in terms of the notation (10.96).

Exercise 10.17 Construct the solution of the Cauchy problem (10.98) with homogeneous boundary conditions by means of Duhamel's principle. Compare the result with that of the exercise at the end of Sect. 10.10.

Exercise 10.18 Apply Duhamel's principle to the general inhomogeneous heat equation over a finite domain, when both the boundary and the initial conditions are zero. For the solution of the associated homogeneous problem, use the eigenfunction expansion method. Show that the result is identical to that obtained by expanding the forcing function in terms of eigenfunctions, rather than using Duhamel's principle.

References

1. Courant R, Hilbert D (1962) Methods of mathematical physics, vol I. Interscience, Wiley, New York
2. Epstein M (2012) The elements of continuum biomechanics. Wiley, London
3. Farlow SJ (1993) Partial differential equations for scientists and engineers. Dover, New York
4. John F (1982) Partial differential equations. Springer, Berlin
5. Petrovsky IG (1954) Lectures on partial differential equations. Interscience, New York (Reprinted by Dover (1991))
6. Zauderer E (1998) Partial differential equations of applied mathematics, 2nd edn. Interscience, Wiley, New York

Chapter 11
The Laplace Equation

The Laplace equation is the archetypal elliptic equation. It appears in many applications when studying the steady state of physical systems that are otherwise governed by hyperbolic or parabolic operators. Correspondingly, elliptic equations require the specification of boundary data only, and the Cauchy (initial-value) problem does not arise. The boundary data of a second-order elliptic operator offer a choice between two extremes: either the function or its transverse derivative must be specified at the boundary, but not both independently. From the physical standpoint, this dichotomy makes perfect sense. Thus, in a body in equilibrium, one expects to specify a support displacement or the associated reaction force, but not both.

11.1 Introduction

The Laplace equation

$$u_{xx} + u_{yy} = 0, \tag{11.1}$$

for a scalar function $u = u(x, y)$, and its inhomogeneous version

$$u_{xx} + u_{yy} = f(x, y), \tag{11.2}$$

also known as the *Poisson equation*, as well as their three-dimensional counterparts

$$u_{xx} + u_{yy} + u_{zz} = 0, \tag{11.3}$$

and

$$u_{xx} + u_{yy} + u_{zz} = f(x, y, z), \tag{11.4}$$

M. Epstein, *Partial Differential Equations*, Mathematical Engineering,
DOI 10.1007/978-3-319-55212-5_11

are ubiquitous in Physics and Engineering applications. We have already mentioned their appearance in the theory of static elasticity for the description of the small transverse deflections of a tensed membrane. Another application in elasticity is the torsion of a shaft of non-circular cross section. In fluid mechanics, the Laplace equation appears in connection with the irrotational motion of a perfect fluid. A similar application pertains to acoustics. In electrostatics, Poisson's equation relates the charge density with the electrostatic potential. In the classical theory of gravitation, Poisson's equation relates the mass density with the gravitational potential. In heat conduction, the Laplace equation appears in connection with the steady state. In general, many hyperbolic equations in two or three spatial dimensions have a spatial part given by the Laplace operator, also called the *Laplacian*, defined (in Cartesian coordinates) as

$$\nabla^2 u = u_{xx} + u_{yy} + u_{zz}. \tag{11.5}$$

If it appears convenient (for reasons of symmetry, for example) to formulate a problem in a curvilinear coordinate system while keeping the same physical meaning for the field variable, then the Laplace operator must change accordingly.

A function satisfying Laplace's equation is said to be *harmonic*.[1] In an n-dimensional setting, the Laplacian is given by the obvious extension of Eq. (11.5) to n independent variables. Equations (11.3) and (11.4) are extended accordingly. The equations of Laplace and Poisson, as we have already remarked in Chap. 6, are of the elliptic type. They possess no characteristic directions and, correspondingly, the nature of the domains of definition and of the boundary conditions that lead to properly formulated problems with unique solutions is different from the case of hyperbolic and parabolic equations.

11.2 Green's Theorem and the Dirichlet and Neumann Problems

One of the fundamental theorems of vector calculus is the *divergence theorem*. It states that the integral of the *divergence* of a (sufficiently smooth) vector field over a (sufficiently regular) domain $\mathcal{D}$ in $\mathbb{R}^n$ is equal to the *flux* of the vector field over the boundary $\partial\mathcal{D}$ of this domain. We recall that, in Cartesian coordinates $x_1, \ldots, x_n$, the divergence of a vector field $\mathbf{v}$ with components $v_1, \ldots, v_n$, is defined as

$$\operatorname{div} \mathbf{v} = \frac{\partial v_1}{\partial x_1} + \frac{\partial v_2}{\partial x_2} + \ldots + \frac{\partial v_n}{\partial x_n}. \tag{11.6}$$

The flux of the vector $\mathbf{v}$ over an oriented surface element dA with (exterior) unit normal $\mathbf{n}$, is obtained by projecting the vector on the normal and multiplying by the

[1] Notice that we have also used the term harmonic to designate a sinusoidal function of one variable. These two usages are unrelated.

area of the element. In terms of these definitions, therefore, the divergence theorem establishes that

$$\int_{\mathcal{D}} (\text{div}\, \mathbf{v})\, dV = \int_{\partial\mathcal{D}} \mathbf{v} \cdot \mathbf{n}\, dA. \tag{11.7}$$

We will use this theorem (whose proof can be found in any good textbook of Calculus) to derive some useful expressions involving the Laplacian operator.

Consider a differentiable function $u(x_1, ..., x_n)$. The gradient of this function is the vector field ∇u with components

$$\{\nabla u\} = \begin{Bmatrix} \frac{\partial u}{\partial x_1} \\ \cdot \\ \cdot \\ \cdot \\ \frac{\partial u}{\partial x_n} \end{Bmatrix} \tag{11.8}$$

The divergence of the gradient is, therefore, precisely the Laplacian. We conclude that the integral of the Laplacian of a scalar field over a domain is equal to the flux of its gradient through the boundary. In this case, the dot product of the vector field (i.e., the gradient) with the exterior unit normal is the directional derivative of the scalar field in the exterior normal direction $\mathbf{n}$. Thus, we can write

$$\int_{\mathcal{D}} \nabla^2 u\, dV = \int_{\partial\mathcal{D}} \frac{du}{dn}\, dA. \tag{11.9}$$

Consider next two scalar fields u, v. Applying the divergence theorem to the product of one field times the gradient of the other, we obtain

$$\int_{\partial\mathcal{D}} v \frac{du}{dn}\, dA = \int_{\mathcal{D}} \left(v\nabla^2 u + \nabla u \cdot \nabla v\right)\, dV. \tag{11.10}$$

Subtracting the expression obtained by interchanging the fields, yields

$$\int_{\partial\mathcal{D}} \left(v\frac{du}{dn} - u\frac{dv}{dn}\right)\, dA = \int_{\mathcal{D}} \left(v\nabla^2 u - u\nabla^2 v\right)\, dV. \tag{11.11}$$

Equations (11.10) and (11.11) are known as Green's identities.

For the particular case in which both scalar fields are made to coincide, Eq. (11.10) yields the result

$$\int_{\partial\mathcal{D}} u \frac{du}{dn}\, dA = \int_{\mathcal{D}} \left(u\nabla^2 u + \nabla u \cdot \nabla u\right)\, dV. \tag{11.12}$$

Suppose now that the function u is harmonic over the domain $\mathcal{D}$ and that it vanishes on its boundary $\partial\mathcal{D}$. From Eq. (11.12) it will follow that the integral of the square of the magnitude of the gradient must vanish. But for a continuous and non-negative function this is possible only if the gradient vanishes identically within the domain. In other words, all its partial derivatives vanish identically, so that the function must, in fact, be a constant. But since the function has been assumed to vanish over the boundary of the domain, we conclude that it must vanish over the whole of $\mathcal{D}$. This rather trivial observation implies the uniqueness of the solution of the so-called *Dirichlet problem* defined as follows.

Dirichlet problem: Find a solution u of the Poisson equation $\nabla^2 u = f$ over a domain $\mathcal{D}$ with prescribed values of u on the boundary $\partial\mathcal{D}$. The proof of uniqueness[2] is straightforward. Indeed, assume that there exist two solutions to this problem. Since the Poisson equation is linear, the difference between these two solutions must be harmonic (i.e., must satisfy the Laplace equation) and attain a zero value over the whole boundary. It follows from our previous reasoning that this difference must be identically zero, so that both solutions coincide.

Consider now the case in which u is harmonic over the domain $\mathcal{D}$ and that its normal derivative (rather than the function itself) vanishes on the boundary $\partial\mathcal{D}$. Again, by applying Eq. (11.12), we arrive at the conclusion that u must be a constant. Nevertheless, in this case, we can no longer conclude that it must vanish. We thus obtain the following statement about the solution of the so-called *Neumann problem*.[3]

Neumann problem: Find a solution of the Poisson equation $\nabla^2 u = f$ over a domain $\mathcal{D}$ with prescribed values of the normal derivative on $\partial\mathcal{D}$. A solution of this problem is determined uniquely to within an additive constant. Moreover, according to Eq. (11.9), the solution can only exist if the boundary data satisfy the auxiliary condition

$$\int_{\partial\mathcal{D}} \frac{du}{dn}\, dA = \int_{\mathcal{D}} f\, dV. \tag{11.13}$$

Remark 11.1 Intuitively, we can imagine that the Dirichlet problem corresponds to the specification of displacements, while the Neumann problem corresponds to the specification of boundary tractions in an elastic structure. This explains why the Neumann problem requires the satisfaction of an auxiliary condition: The tractions must be in equilibrium with the applied body forces. If they are not, a solution cannot exist within the realm of statics. The dynamic problem is, of course, governed by hyperbolic equations, such as the wave equation, which necessitate the specification of initial displacements and velocities and which include the forces of inertia.

Notice that, since the specification of the value of the solution on the boundary is enough to determine a unique solution (if such a solution indeed exists), we cannot

---

[2]Strictly speaking, this proof of uniqueness requires the solution to be twice differentiable not just in the interior but also on the boundary of the domain. This requirement can be relaxed if the proof is based on the maximum-minimum principle, that we shall study below.

[3]Named after the German mathematician Carl Gottfried Neumann (1832–1925), not to be confused with John von Neumann (1903–1957), the great Hungarian-American mathematician.

simultaneously specify both the function and its normal derivative on the boundary. In other words, the Cauchy problem for the Laplace equation has no solution in general. This fact is in marked contrast with hyperbolic equations. It can be shown that the Cauchy problem for the Laplace equation is in general unsolvable even locally.[4] We have already seen that a solution of the heat equation must be of class C^∞. In the case of Laplace's equation, the complete absence of characteristic directions, leads one to guess that perhaps this will also be the case, since no discontinuities can be tolerated. It can be shown, in fact, that bounded solutions of Laplace's equation must be not just C^∞, but also (real) analytic (i.e., they must have convergent Taylor-series expansions in an open neighbourhood of every point).

11.3 The Maximum-Minimum Principle

Theorem 11.1 (Maximum-minimum theorem)[5] *A harmonic function which is continuous in $\mathcal{D} \cup \partial\mathcal{D}$ (namely, in the union of the interior of a bounded domain and its boundary) attains its maximum and minimum values on the boundary $\partial\mathcal{D}$.*

Proof The proof can be carried out along similar lines as in the case of the parabolic equation. Since the boundary $\partial\mathcal{D}$ is a closed and bounded set, the restriction of u to $\partial\mathcal{D}$ must attain a maximum value m at some point of $\partial\mathcal{D}$. On the other hand, since $\mathcal{D} \cup \partial\mathcal{D}$ is also closed and bounded, the function must attain its maximum M at some point P of $\mathcal{D} \cup \partial\mathcal{D}$. Let us assume that P is an interior point and that, moreover, $M > m$. Without any loss of generality, we may assume that the origin of coordinates is at P. Let us now construct the auxiliary function

$$v = u + \frac{M-m}{2d^2}\, r^2. \tag{11.14}$$

In this expression, r denotes the length of the position vector and d is the *diameter* of $\mathcal{D}$.[6] This function v is strictly larger than u, except at P, where they have the same value, namely M. The restriction of v to $\partial\mathcal{D}$, on the other hand, will satisfy the inequality

$$v \leq m + \frac{M-m}{2} < M. \tag{11.15}$$

We conclude that v attains its maximum at an interior point. On the other hand,

[4]See [3], p. 98. Notice that, correspondingly, the Dirichlet problem has in general no solution for the hyperbolic and parabolic cases, since the specification of the solution over the whole boundary of a domain will in general lead to a contradiction. For this point see [2], p. 236.

[5]See [4], p. 169.

[6]Recall that the diameter of a set (in a metric space) is the least upper bound of the distances between all pairs of points of the set.

$$\nabla^2 v = \nabla^2 u + \frac{M-m}{d^2} = \frac{M-m}{d^2} > 0. \tag{11.16}$$

Since, however, at an interior point of a domain any maximum of a differentiable function must be a relative maximum, none of the second partial derivatives can be positive, making the satisfaction of Eq. (11.16) impossible. Having arrived at this contradiction, we conclude that the maximum of u is attained at the boundary $\partial\mathcal{D}$. Changing u to $-u$, we can prove that the same is true for the minimum value. □

Just as in the case of the parabolic heat equation, we can prove as corollaries of the maximum-minimum theorem the uniqueness and continuous dependence on the boundary data of the Dirichlet problem.

A nice intuitive visualization of the maximum-minimum principle can be gathered from the case of a membrane (or a soap film) extended over a rigid closed frame. If we give the initially plane frame a small transverse deformation (i.e., a warping), we do not expect the membrane to bulge either upwards or downwards beyond the frame, unless external forces are applied. A similar intuitive interpretation can be stated in the realm of the thermal steady state over some plane region. The temperature attains its maximum and minimum values at the boundaries of the region.

11.4 The Fundamental Solutions

There are many ways to tackle the difficult problem of solving the Laplace and Poisson equations in some degree of generality. A useful result towards this end is obtained by investigating the possible existence of spherically symmetric solutions. Let $P \in \mathbb{R}^n$ be a point with coordinates $\bar{x}_1, ..., \bar{x}_n$ and let $r = r(x_1, ..., x_n)$ denote the distance function to P, namely,

$$r = +\sqrt{\sum_{j=1}^{n} \left(x_j - \bar{x}_j\right)^2}. \tag{11.17}$$

A function is said to be *spherically symmetric* with respect to P if it can be expressed as a function of the single variable r, viz.,

$$u = g(r). \tag{11.18}$$

If this is the case, it is not difficult to calculate its Laplacian. Indeed, denoting by primes the derivatives of g with respect to its independent variable r, we obtain

$$\frac{\partial u}{\partial x_k} = g' \frac{\partial r}{\partial x_k} = g' \frac{x_k - \bar{x}_k}{r}. \tag{11.19}$$

Similarly, for the second derivatives we get

$$\frac{\partial^2 u}{\partial x_m \partial x_k} = g'' \frac{(x_m - \bar{x}_m)(x_k - \bar{x}_k)}{r^2} + g' \frac{r^2 \delta_{km} - (x_m - \bar{x}_m)(x_k - \bar{x}_k)}{r^3} \tag{11.20}$$

Making $m = k$ and adding for $k = 1, ..., n$ yields the desired result as

$$\nabla^2 u = g'' \frac{n-1}{r} g'. \tag{11.21}$$

If we wish to satisfy Laplace's equation, therefore, we are led to the solution of a simple linear ODE. Specifically,

$$g'' \frac{n-1}{r} g' = 0. \tag{11.22}$$

The solution, which exists on the whole of $\mathbb{R}^n$, except at P (where it becomes unbounded), is given by

$$g = \begin{cases} A + \frac{B}{r^{n-2}} & \text{if } n > 2 \\ A + B \ln r & \text{if } n = 2 \end{cases}, \tag{11.23}$$

where A and B are arbitrary constants. Notice that in the particularly important case $n = 3$ the solution is a linear function of the reciprocal distance (from the physical point of view, this corresponds to the electrostatic or gravitational potentials of a concentrated charge or mass).

To reveal the meaning of the solution (11.23), let us consider the following (three-dimensional) problem associated with a sphere of radius ε with centre at P. We look for a bounded C^1 spherically-symmetric function $u(x, y, z)$ vanishing at infinity and satisfying the Poisson equation

$$\nabla^2 u = \frac{1}{\frac{4}{3}\pi\, \varepsilon^2 r} \qquad r \leq \varepsilon, \tag{11.24}$$

and the Laplace equation

$$\nabla^2 u = 0 \qquad r > \varepsilon. \tag{11.25}$$

Just as before, the assumed spherical symmetry allows us to reduce this problem to that of an ODE, namely, setting $u = g(r; \varepsilon)$,

$$g''(r; \varepsilon) + \frac{2}{r} g'(r; \varepsilon) = \begin{cases} \frac{3}{4\pi\, \varepsilon^3} & \text{if } r \leq \varepsilon \\ 0 & \text{if } r > \varepsilon \end{cases}. \tag{11.26}$$

The solution of this problem is easily obtained as

$$g(r;\varepsilon) = \begin{cases} \frac{1}{8\pi\,\varepsilon}\left(\frac{r^2}{\varepsilon^2} - 3\right) & \text{if } r \le \varepsilon \\ -\frac{1}{4\pi\, r} & \text{if } r > \varepsilon \end{cases}. \tag{11.27}$$

A remarkable feature of the solution just found is that the solution outside the sphere is independent of the size of the sphere.[7] As the radius of the sphere tends to zero, the right-hand side of Eq. (11.25) approaches δ_P, the (three-dimensional) Dirac delta function at P. This means that, with $A = 0$ and with the value of B appropriately calibrated (for each dimension), Eq. (11.23) provides the solution of the problem

$$\nabla^2 u = \delta_P, \tag{11.28}$$

with zero boundary condition at infinity. We call this a *fundamental solution* of the Laplace equation with *pole* P. This solution depends both on the coordinates x_j of the variable point in space and the coordinates $\bar{x}_j$ of point P. It is sometimes convenient to express this double dependence explicitly with the notation $K(x_j, \bar{x}_j)$. The explicit formulas for the fundamental solution in an arbitrary dimension can be obtained in a similar way.[8]

Any solution of Eq. (11.28) in some domain $\mathcal{D} \subset \mathbb{R}^n$ containing P, regardless of boundary conditions, will also be called a fundamental solution with pole P. Clearly, if w is harmonic within this domain, the new function

$$G(x_j, \bar{x}_j) = K(x_j, \bar{x}_j) + w(x_j) \tag{11.29}$$

is also a fundamental solution with pole P.

11.5 Green's Functions

A fundamental solution that satisfies zero boundary conditions on the boundary of a given domain is called *Green's function* for that domain. The importance of Green's functions (also known as *influence functions*, for obvious reasons) is that the solution of the Poisson equation

$$\nabla^2 u = f(x_j) \tag{11.30}$$

on a domain $\mathcal{D}$ with u vanishing at the boundary is given by the superposition integral

[7] This feature of the solution for the case of the gravitational field was extremely important to Newton, who was at pains to prove it. It is this property that allowed him to conclude that the forces exerted by a homogeneous sphere on empty space are unchanged if the total mass is concentrated at its centre.

[8] For the case $n = 2$, we have $B = 1/2\pi$. For $n = 3$, as we have seen, the value is $B = -1/4\pi$. For higher dimensions, the value of the constant can be shown to be related to the 'area' of the corresponding hyper-sphere, which involves the Gamma function. See [3], p. 96.

$$u(x_j) = \int_{\mathcal{D}} G(x_j, \xi_j)\, f(\xi_j)\, dV_\xi. \tag{11.31}$$

The solution of the Dirichlet problem for the Laplace equation with inhomogeneous boundary conditions, can be obtained in a similar way. Indeed, let the boundary values be given by

$$u|_{\partial\mathcal{D}} = h(x_j). \tag{11.32}$$

Let us assume that this function h has enough smoothness that we can extend it (non-uniquely, of course) to a C^∞ function $\hat{h}$ defined over the whole domain $\mathcal{D}$. Then clearly the function

$$v = u - \hat{h} \tag{11.33}$$

satisfies, by construction, homogeneous boundary conditions, but the inhomogeneous (Poisson) equation

$$\nabla^2 v = -\nabla^2 \hat{h}. \tag{11.34}$$

In terms of the Green function (assumed to be known for the domain under consideration), the solution of this problem is given by Eq. (11.31) as

$$v(x_j) = -\int_{\mathcal{D}} G(x_j, \xi_j)\, \nabla^2 \hat{h}(\xi_j)\, dV_\xi. \tag{11.35}$$

From Eq. (11.33) we obtain the solution of the original Dirichlet problem as

$$u(x_j) = \hat{h}(x_j) - \int_{\mathcal{D}} G(x_j, \xi_j)\, \nabla^2 \hat{h}(\xi_j)\, dV_\xi. \tag{11.36}$$

This expression would seem to indicate that the solution depends on the particular extension $\hat{h}$ adopted. That this is not the case can be deduced by a straightforward application of Eq. (11.11), appropriately called a Green identity, which yields

$$\int_{\mathcal{D}} G(x_j, \xi_j)\, \nabla^2 \hat{h}(\xi_j)\, dV_\xi = \hat{h}(x_j) - \int_{\partial\mathcal{D}} \left(h(\xi_j) \frac{dG(x_j, \xi_j)}{dn} \right) dA_\xi. \tag{11.37}$$

In obtaining this result, the properties of each of the functions involved were exploited. Combining Eqs. (11.36) and (11.37), we obtain the final result

$$u(x_j) = \int_{\partial\mathcal{D}} \left(h(\xi_j) \frac{dG(x_j, \xi_j)}{dn} \right) dA_\xi \tag{11.38}$$

Thus, the solution involves only the values of the data at the boundary, rather than any extension to the interior. Expression (11.41) is regular at every interior point of the domain.

Although we have not provided rigorous proofs of any of the preceding theorems (proofs that can be found in the specialized books),[9] we have attempted to present enough circumstantial evidence to make these results at least plausible. The main conclusion so far is that to solve a Dirichlet problem over a given domain, whether for the Laplace or the Poisson equation, can be considered equivalent to finding Green's function for that domain. It is important to realize, however, that finding Green's function is itself a Dirichlet problem.

11.6 The Mean-Value Theorem

An important result that characterizes harmonic functions is the following.

Theorem 11.2 (Mean value theorem) *The value of a harmonic function u at any point is equal to the average of its values over the surface of any sphere with centre at that point.*

Proof The theorem is valid for the circle, the sphere or, in the general case, an n-dimensional ball. For specificity, we will consider the (three-dimensional) case of a sphere. Let P be a point with coordinates $\hat{x}_j$ in the domain under consideration and let $\mathcal{B}$ denote the (solid) sphere with centre P and radius R. Always keeping the centre fixed, we proceed to apply Green's formula (11.11) to the fundamental solution $K(x_j, \bar{x}_j)$ and to the harmonic function $u(x_j)$ and we notice the vanishing of the term $\nabla^2 u$ (by hypothesis), and obtain

$$\int_{\mathcal{B}} u \, \nabla^2 K \, dV = \int_{\partial\mathcal{B}} \left(u \frac{dK}{dn} - K \frac{du}{dn} \right) dA. \tag{11.39}$$

But, due to the spherical symmetry of K, both it and its derivative in the normal direction (which is clearly radial) are constant over the boundary of the ball. Recalling that, by a direct application of the divergence theorem, the flux of the gradient of a harmonic function vanishes over the boundary of any domain, we conclude that the last term under the right-hand side integral vanishes. The normal derivative of K at the boundary is obtained directly from Eq. (11.23), or (11.27), as

$$\frac{dK}{dn} = \frac{dK}{dr} = \frac{1}{4\pi R^2}. \tag{11.40}$$

Finally, invoking Eq. (11.28) for the fundamental solution K, we obtain

---

[9]Good sources are [1, 3, 5].

$$u(\hat{x}_j) = \frac{1}{4\pi R^2} \int\limits_{\partial\mathcal{B}} u \, dA, \tag{11.41}$$

which is the desired result. □

Corollary 11.1 *The value of a harmonic function at a point is also equal to the volume average over any ball with centre at that point.*

Proof Trivial. □

Remark 11.2 It is remarkable that the converse of the mean value theorem also holds. More specifically, if a continuous function over a given domain satisfies the mean value formula for every ball contained in this domain, then this function is harmonic in the domain. For a rigorous proof of this fact, see [1], p. 277.

11.7 Green's Function for the Circle and the Sphere

It is, in fact, not difficult to construct explicit Green's functions for a domain which is a circle, a sphere or, in the general case, an n-dimensional ball of given radius R. The construction is based on a simple geometric property of circles and spheres. We digress briefly to show this property in the case of a circle (the extension to three dimensions is rather obvious, by rotational symmetry).

Given a circle of centre P and radius R, as shown in Fig. 11.1, let $Q \neq P$ be an arbitrary internal point. Extending the radius through Q, we place on this line another point S outside the circle, called the it reflected image of Q, according to the proportion

$$\frac{\overline{PS}}{\overline{PQ}} = \left(\frac{R}{\overline{PQ}}\right)^2. \tag{11.42}$$

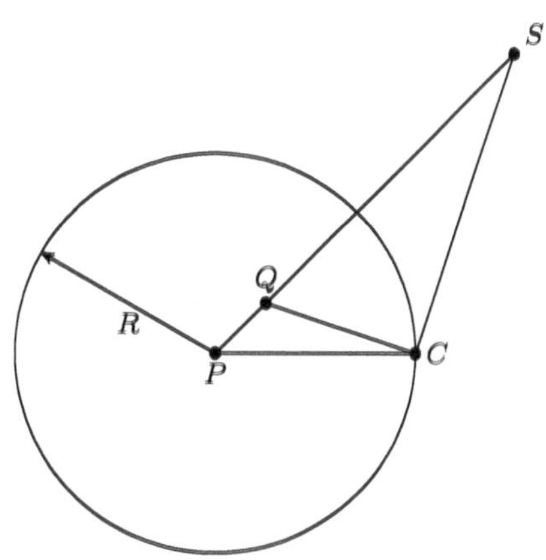

Fig. 11.1 Green's function argument

It is clear that this point lies outside the circle. Let C be a point on the circumference. The triangles QPC and CPS are similar, since they share the angle at P and the ratio of the adjacent sides. Indeed, by Eq. (11.42),

$$\frac{\overline{PS}}{\overline{PC}} = \frac{\overline{PC}}{\overline{PQ}}. \tag{11.43}$$

It follows that the ratio of the remaining sides must be the same, namely,

$$\frac{\overline{CS}}{\overline{CQ}} = \frac{\overline{PC}}{\overline{PQ}}. \tag{11.44}$$

The right-hand side of this equation is independent on the particular point C chosen on the circumference. It depends only on the radius of the circle and the radial distance to the fixed point Q.

We use this property to construct Green's function for the circle (or the sphere) $\mathcal{B}$. We start by noting that (always keeping Q and, therefore, also S fixed) the fundamental solution $K(X, S)$, where X denotes an arbitrary point belonging to $\mathcal{B}$, is smooth in $\mathcal{B}$ and its boundary $\partial\mathcal{B}$. This follows from the fact that S is an exterior point. Moreover, if the point X happens to belong to the boundary, the value of $K(X, S)$ is given by

$$K(X, S)|_{X\in\partial\mathcal{B}} = \left(\frac{\overline{PQ}}{R}\right)^{n-2} K(X, Q)|_{X\in\partial\mathcal{B}}\,. \tag{11.45}$$

This result is a direct consequence of Eq. (11.44) and the general formula (11.23) for $n > 2$. For $n = 2$, a similar (logarithmic) formula applies. Therefore, the function

$$G(X, Q) = K(X, Q) - \left(\frac{\overline{PQ}}{R}\right)^{2-n} K(X, S), \tag{11.46}$$

is Green's function for the ball.

For the circle ($n = 2$), the corresponding formula is

$$G(X, Q) = K(X, Q) - K(X, S) - \frac{1}{2\pi}\ln\left(\frac{\overline{PQ}}{R}\right). \tag{11.47}$$

Assume that a Dirichlet problem has been given by specifying the value of the field on the boundary as a function $h(Y)$, $Y \in \partial\mathcal{B}$. According to Eq. (11.41), the solution of this problem is given by

$$u(X) = \int\limits_{\partial\mathcal{B}} \left(h(Y)\,\frac{dG(X, Y)}{dn}\right) dA_y. \tag{11.48}$$

We need, therefore, to calculate the normal (i.e., radial) derivative of Green's functions (just derived) at the boundary of the ball. When this is done carefully, the result is

$$\frac{dG(X,Y)}{dn_Y} = H(X,Y) = \frac{1}{4\pi R}\,\frac{R^2 - \overline{PX}^2}{\overline{XY}^2}. \tag{11.49}$$

This is the formula for the sphere. For the circle, the denominator has a 2 rather than a 4. Equation (11.48), with the use of (11.49), is known as *Poisson's formula*. It solves the general Dirichlet problem for a ball.

Exercises

Exercise 11.1 Show that under an orthogonal change of coordinates, the Laplacian retains the form given in Eq. (11.5). If you prefer to do so, work in a two-dimensional setting.

Exercise 11.2 Express the Laplacian in cylindrical coordinates (or polar coordinates, if you prefer to work in two dimensions).

Exercise 11.3 Obtain Eqs. (11.9), (11.10) and (11.11). Show, moreover, that the flux of the gradient of a harmonic function over the boundary of any bounded domain vanishes.

Exercise 11.4 A membrane is extended between two horizontal concentric rigid rings of 500 and 50 mm radii. If the inner ring is displaced vertically upward by an amount of 100 mm, find the resulting shape of the membrane. Hint: Use Eq. (11.23).

Exercise 11.5 Carry out the calculations leading to (11.27). Make sure to use each of the assumptions made about the solution. Find the solution for the two-dimensional case.

Exercise 11.6 Carry out and justify all the steps necessary to obtain Eq. (11.38).

Exercise 11.7 What is the value of $G(X, P)$? How is this value reconciled with Eq. (11.46)?

Exercise 11.8 Adopting polar coordinate in the plane, with origin at the centre of the circle, and denoting the polar coordinates of X by ρ, θ and those of Y (at the boundary) by R, ψ, show that the solution to Dirichlet's problem is given by the expression

$$u(\rho,\theta) = \frac{1}{2\pi}\int_0^{2\pi} \frac{R^2 - \rho^2}{R^2 + \rho^2 - 2R\,\rho\,\cos(\theta - \psi)}\,h(\psi)\,d\psi. \tag{11.50}$$

where $h(\psi)$ represents the boundary data. Use Eqs. (11.48) and (11.49) (with a 2 in the denominator).

References

1. Courant R, Hilbert D (1962) Methods of mathematical physics, vol II. Interscience, Wiley, New York
2. Garabedian PR (1964) Partial differential equations. Wiley, New York
3. John F (1982) Partial differential equations. Springer, Berlin
4. Petrovsky IG (1991) Lectures on partial differential equations. Dover, New York
5. Sobolev SL (1989) Partial differential equations of mathematical physics. Dover, New York

Index

M. Epstein, *Partial Differential Equations*, Mathematical Engineering,
DOI 10.1007/978-3-319-55212-5

MIX
Papier aus verantwortungsvollen Quellen
Paper from responsible sources
FSC® C105338

If you have any concerns about our products,
you can contact us on
ProductSafety@springernature.com

In case Publisher is established outside the EU,
the EU authorized representative is:
Springer Nature Customer Service Center GmbH
Europaplatz 3, 69115 Heidelberg, Germany

Printed by Libri Plureos GmbH
in Hamburg, Germany